Nancy Segal Janes

Cuisenaire Company of America, Inc.
White Plains, New York

Managing editor: Doris Hirschhorn
Development editor: Jane O'Sullivan
Design manager: Phyllis Aycock
Cover design: Tracey Munz
Production: Tracey Munz

PO Box 5026
White Plains, New York 10602-5026

Printed in the United States of America.

ISBN 0-938587-96-X

1 2 3 4 5 00 99 98 97 96

ACKNOWLEDGMENTS

Many of the games and puzzles in this book have appeared in *Wonderful Ideas*, a newsletter featuring ideas for teaching, learning, and enjoying mathematics. Over the years, many teachers and students have contributed their wonderful ideas toward the development of these activities. Although they are too numerous to thank individually, I would like to acknowledge, in particular, Rachel McAnallen and the Institute for Math Mania, Chris Kauth and his students at the Buckingham, Browne & Nichols School, Doris Hirschhorn, and Jane O'Sullivan. And, a special thanks to George and Sophia.

CONTENTS

TEACHER NOTES

RECORDING SHEETS

INTRODUCTION

"The having of wonderful ideas is what I consider
the essence of intellectual development."
—Eleanor Duckworth

I had been teaching math to elementary and middle-school students for about seven years when, in graduate school, I first encountered Eleanor Duckworth and her notion of wonderful ideas. To Eleanor Duckworth, wonderful ideas are any ideas that come from students, developed through active involvement with their learning. The ideas do not have to be new or unique. What is important is that the young learners created these ideas themselves. I remembered all the "Aha!" moments that occurred in my classroom when students finally solved a challenging problem, and I knew that those were wonderful ideas for those students. As I thought more about this, I realized that these "Aha!" incidents were most likely to occur during hands-on, out-of-the-ordinary, problem-solving situations. In my classroom (and in my work with teachers), I have always tried to give learners the opportunity to explore, uncover, and understand concepts and ideas on their own before we discuss them.

With this vision in mind, I created *Counter Logic*, a collection of problem-solving games and puzzles that enable students to play around with and reason with mathematical ideas as they try to uncover patterns, strategies, and solutions. The activities are intended to be fun and intriguing, but also challenging and thought-provoking. As you try the activities in the book, you may hear many wonderful ideas from your learners: "I got it!" "Oh, I see a different pattern!" or maybe even "I didn't notice that the first time I played." Encourage your students to explore and explain their wonderful ideas, and you may discover that their wonderful ideas lead to more wonderful ideas for others.

ORGANIZATION OF THE BOOK

Counter Logic has three parts: student activities, teacher notes, and recording sheets.

Student Activities

The first part of the book is for the student. It contains games and puzzles, most of which are matched with a game board on the facing page. Each student page is divided into three sections:

You Need lists the materials needed for the activity.

How to Play provides information on the way to play the game or the puzzle, explains what constitutes a win, and often shows sample turns or moves and a sample solution.

Working Together offers suggestions for working with a partner or in small groups to discuss such things as strategies or patterns and ways to record solutions for the activities.

The activities are divided into two general groups: Puzzles, which can be solved by a student working alone or in a small group, and Strategy Games, which require two players. Within each group, the activities are placed in approximate order, beginning with simpler activities and working toward the more challenging ones. However, some of the games and variations throughout may be challenging (or easy) for different groups of students. The activities can be used in any order, depending on the interests and abilities of your class.

Teachers Notes

The second part of the book contains teacher notes for each activity. These notes consist of a brief overview of each activity, solutions, and strategies for solving the puzzles or winning the games. Also included are variations or extensions that will enable students and teachers to delve deeper into the mathematics and critical thinking behind each activity. Many of the variations are more challenging than the original activity, but some are simpler, particularly suited for students who are having difficulty or find themselves in need of a hint to set them off in the right direction. These variations, in fact, and the Working Together sections from the student pages can, in some cases, serve as homework assignments or as family math activities.

Teachers should keep in mind that students may and often will come up with solutions and strategies different from the ones presented in the teacher notes. (These may be some of the wonderful ideas mentioned earlier!) The solutions and strategies described in the book are intended to provide only a beginning for what might ensue in a classroom discussion. By no means expect all students to come up with the same strategies presented, and be sure to encourage students to explain their unique solutions.

Recording Sheets

The final section of the book, beginning on page 67, has recording sheets for some of the activities that involve more complicated diagrams. These can be duplicated as needed.

Materials

Transparent counters are suggested because they come in five colors, can be used on an overhead projector, and allow students to see the numbers or words that are covered. Any type of counter, however, can be used. Use whatever is handy: colored chips, raisins, beans, sugar cubes, cardboard squares or circles—any small game pieces will do. Use counters that will work well with your students. Raisins are a nice, inexpensive material, but may disappear quickly if used with young or hungry students! A few activities call for different-colored counters, but any distinct counters are fine.

CLASSROOM MANAGEMENT SUGGESTIONS

Trying any new activity or using any new material in the classroom for the first time involves some management issues for teachers. I recommend playing the games beforehand so that you can begin to think about the strategies, patterns, and thinking processes involved in the activity. If you do this, then you can use the in-class time to observe students and listen to their conversations as they work through strategies and challenging problems. New questions for the Working Together section may occur to you while observing your students. Add these questions to the discussion even if you do not know the answer—sometimes the best questions are the ones with unknown answers.

Allow students plenty of time to play the strategy games. As they play the games over and over again, they will begin to see patterns in the play that will help them formulate strategies.

For most of the *Counter Logic* activities, students can work through the problems with a partner. Be sure, though, to give all students their own copy of the activity or game board so that they can manipulate the counters and work out some solutions on their own. For a solitaire puzzle like Criss-Cross, students can compare and discuss strategies after playing the game by themselves. Interacting in this fashion encourages students to discuss and reflect on their strategies and ideas, a wonderful way for students to begin communicating mathematically.

Many of the activities work well for math stations, where students can try the activity independently during free time or when they have finished another activity early.

MATHEMATICAL NOTES

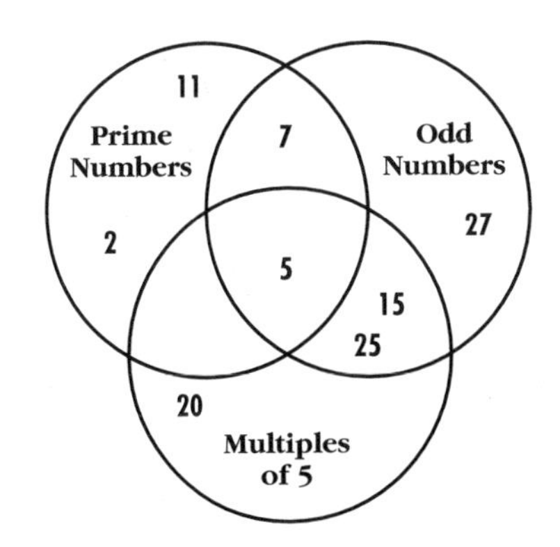

The idea behind Venn diagrams comes into play in several games, such as Ring Teasers, Crazy Cups, and Odd Even Shapes. Venn diagrams are used by mathematicians to explore the idea of sets and subsets. Venn diagrams are a type of graph usually displaying overlapping circles (as shown to the right) that are used to show a logical relationship between members of a set. The areas where the circles overlap represent the *intersection* of the sets—the place at which one item is considered part of two or more sets at the same time.

For example, in the Venn diagram seen here, the circles are labeled "Prime Numbers," "Odd Numbers," and "Multiples of 5." Five is odd, prime, and a multiple of five, so it appears in the intersection of all three circles. Fifteen is both odd and a multiple of five, so it appears in the intersection of those two circles. Seven is both odd and prime, so it appears in the intersection of those two circles.

Many of the problems can be solved with an *empty* set, meaning a group (or row or column) containing zero elements. Zero is considered by mathematicians to be an even number. Knowing this can help generate many solutions to some of the problems in this book, such as Odd Even Circles and Ring Teasers.

NCTM STANDARDS

To encourage a math environment in which wonderful ideas can occur, teachers should keep in mind the *Curriculum and Evaluation Standards for School Mathematics*, published in 1989 by the National Council of Teachers of Mathematics. The goal of NCTM, and of math educators in general, as set forth by the *Standards*, is to revise and enhance the math curriculum taught in today's schools. In the book, the NCTM says that the educational goals for students should be:

1. that they learn to value mathematics,
2. that they become confident in their abilities to do mathematics,
3. that they become mathematical problem solvers,
4. that they learn to communicate mathematically, and
5. that they learn to reason mathematically.

A curriculum that promotes problem solving inevitably involves reasoning by asking "Why," "How," and "What if..." questions and by encouraging students to explore and follow through on their ideas in situations where there is more than one solution and more than one way to reach a solution. Most of the activities in this book encourage such reasoning in that they have several possible solutions and can be solved using a variety of strategies and approaches.

To emphasize reasoning and problem solving, teachers must require students to communicate their discoveries, strategies, and thoughts about mathematics. When students explain their strategies for solving a problem, they must clarify their ideas in order to express them in words. The Working Together section provides opportunities for students to explain and discuss their experiences doing the activities.

Students benefit from hearing different methods for solving problems. When students do not understand one classmate's strategy for solving a problem, another student's approach may make more sense to them. As they begin to hear and see different strategies, patterns, and methods used by classmates, they begin to value their own unique approaches to solving problems. In addition, hearing other students' solutions to one puzzle may open up new ways of thinking and broaden their approach to later puzzles.

DISCUSSIONS IN THE CLASSROOM

Discussions in the classroom are an effective way to promote mathematical communication and are an important part of the Working Together sections. Leading a successful classroom discussion can be challenging. Here are a few hints I've picked up over the years.

- Allow students to share their ideas in a small group before the whole-class discussion begins. This enables more students to be involved and provides a chance for them to clarify their ideas before speaking to the entire group. Most of the *Counter Logic* activities can be done by pairs of students, and these pairs can discuss the Working Together questions on their own.
- Some students feel more comfortable speaking to a small group than to the whole class. One option is to have a group choose one student to summarize its response before opening up the question for class discussion.
- After posing a question, give students a minute or so to jot down a few notes in response. I have found that there always seem to be students who raise their hands before I even finish asking a question, but others need a little more time to formulate their thoughts. Always give students enough time to finish their responses, and do not let other students (or yourself) interrupt the speaker.
- Keep in mind that when you ask students a question, it is because you are interested in their responses. After hearing an answer, many teachers repeat a student's exact words. Doing so takes away the importance of the student's thoughts and merely stresses the answer. If you want the answer repeated because you are not sure if everyone has heard or understood, either ask the student to repeat it so that everyone can hear or ask another student to explain it.
- Writing students' names and responses on a large piece of paper can preserve the discussion long after it has been completed. This makes a nice bulletin-board display, gives credit to small groups' or individuals' contributions, and enables you to remind students of important ideas brought up in previous classes.

Eleanor Duckworth writes that "the more we help children to have their wonderful ideas and to feel good about themselves for having them, the more likely it is that they will some day happen upon wonderful ideas that no one else has happened upon before." Wonderful ideas come in many forms: a winning strategy for Bad Apple, a unique solution in Crazy Cups, or a clever and challenging arrangement for a classmate to solve in Leave One. There are some wonderful ideas out in your classroom just waiting to be uncovered. Enjoy!

Nancy Segal Janes

REFERENCES

Duckworth, Eleanor. *"The Having of Wonderful Ideas" and Other Essays on Teaching and Learning.* New York, NY: Teachers College Press, 1987.

National Council of Teachers of Mathematics. *Curriculum and Evaluation Standards for School Mathematics.* Reston, VA: NCTM, 1989.

STUDENT ACTIVITIES

PUZZLES

LEAVE ONE

YOU NEED

- Up to 12 counters
- Game board

HOW TO PLAY

Set up the counters on the game board to match one of the four arrangements shown below. Remove the counters according to the following rules so that 1 counter is left in the center square. Here are the rules:

- Counters can move only by jumping over an adjacent counter into an empty square.
- Counters can jump only in four directions: right, left, up, or down. Diagonal jumps are not allowed.
- When a counter is jumped, remove that counter from the board.

Now play each of the remaining arrangements.

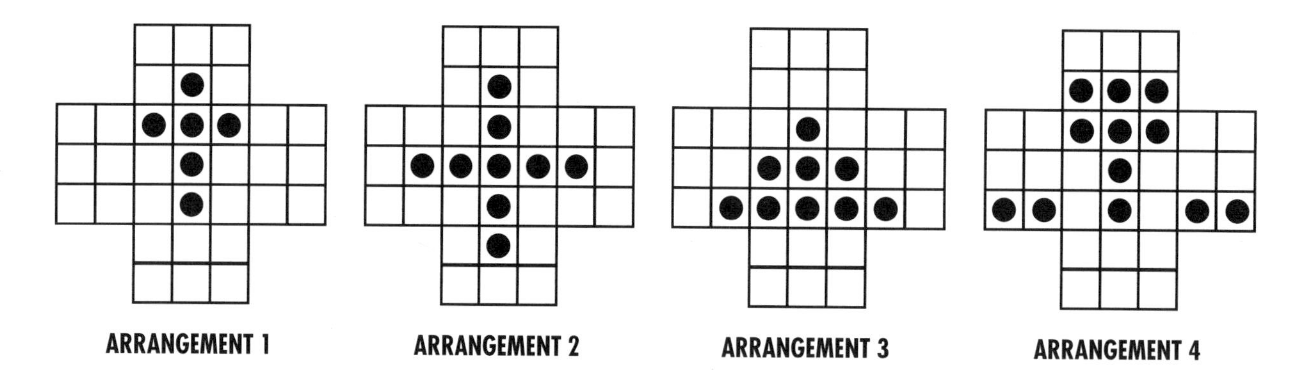

WORKING TOGETHER

- Work with a partner to determine the least number of jumps needed to leave 1 counter in the center of the board for each arrangement.
- Find a way to record the jumps.
- Discuss this: Where should the last 2 counters be placed so that the final jump will leave you with a single counter in the center? the last 3 counters?

LEAVE ONE
GAME BOARD

TIGHT FIT

YOU NEED

- 12 counters
- Game board

HOW TO PLAY

Puzzle 1 Determine the maximum number of counters that can be placed on the game board so that there is no more than 1 counter in any row, column, or diagonal. Record your solution.

Puzzle 2 Determine the maximum number of counters that can be placed on the game board so that there are no more than 2 counters in any row, column, or diagonal. Record your solution.

WORKING TOGETHER

- Work with a partner. Try the same two puzzles on smaller boards, such as a 5 x 5 board or a 4 x 4 board. What changes are there in the maximum number of counters you can place on the board? Describe any patterns you notice.
- Try the puzzles on a larger board, such as a 7 x 7 or 8 x 8 board. Find more than one solution for each puzzle. Record your solutions.

TIGHT FIT

GAME BOARD

COUNTER MOVES

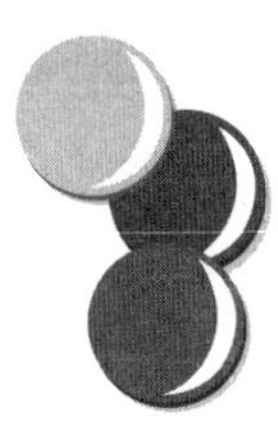

YOU NEED

- Up to 10 counters
- Game board

HOW TO PLAY

Puzzle 1 Set up 8 counters on the game board to form the H-shape shown below. Turn the H-shape into the square shape shown next to it by moving counters according to these rules:

- Counters can be moved only in a straight line (up, down, right, or left), but they can be moved more than one space at a time. A move up and to the right counts as two moves, but a move two spaces to the right counts as only one move.

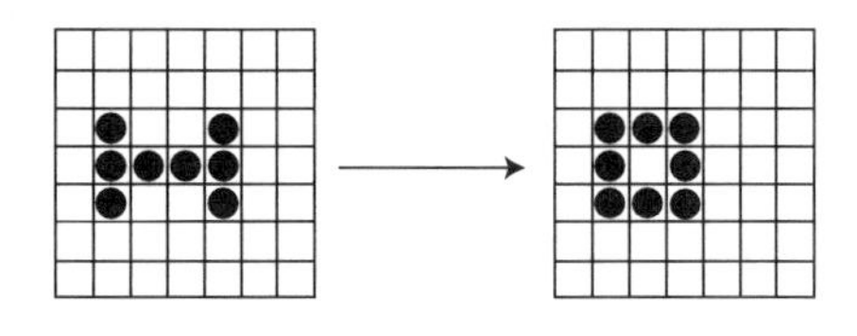

- No diagonal moves are allowed.
- A counter cannot jump over another counter.

Puzzle 2 Set up 7 counters on the game board in the shape shown here. Make ten straight lines of 3 counters each by adding only 2 more counters.

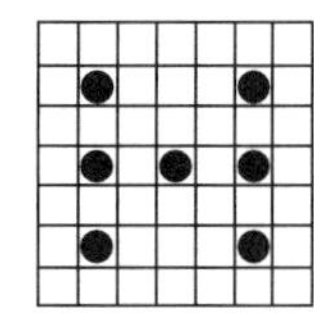

WORKING TOGETHER

Puzzle 1

- Work with a partner to figure out the fewest number of moves needed to turn the H into a square. Record and count your moves.
- Discuss this: What would be the fewest number of moves needed if you were allowed to move diagonally? If you were allowed to jump another counter?

Puzzle 2

- Work with a partner to find a way to record possible solutions.

COUNTER MOVES

GAME BOARD

CIRCLE MYSTERIES, PART I

YOU NEED

- 20 counters
- Game board
- Recording sheets, page 68

HOW TO PLAY

Position counters inside the circles on the game board so that the number in each rectangle and triangle is the same as the total number of counters in the two adjacent circles. For each puzzle:

- Fill in the rectangles and triangles with the given numbers.
- Use the numbers to figure out how many counters to put in each circle.
- If each given number is the sum of the counters in the adjoining two circles, you have found a solution.
- Record your solution.

Here is a sample puzzle and its solution.

Sample

Put these numbers in the rectangles: 5, 6.

Put these numbers in the triangles: 4, 7.

The solution requires 11 counters. Notice that the sum of the counters in Circles 1 and 2 is 5, the number in Rectangle A; the sum of the counters in Circles 1 and 3 is 7, the number in Triangle A; the sum of the counters in Circles 2 and 4 is 4, the number in Triangle B; and, the sum of the counters in Circles 3 and 4 is 6, the number in Rectangle B.

Solution

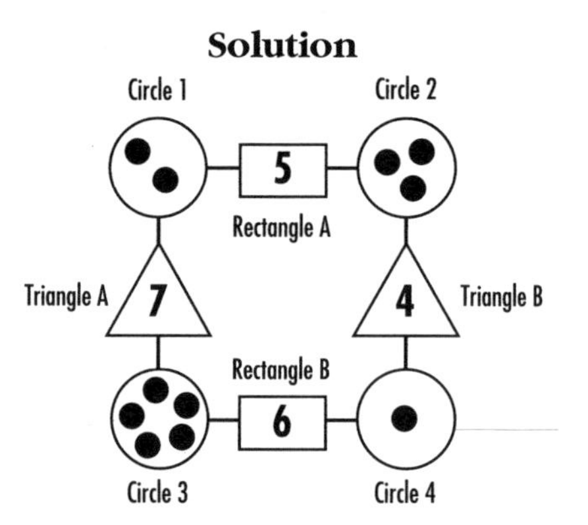

	Numbers for the rectangles	Numbers for the triangles
Puzzle 1	12, 9	14, 7
Puzzle 2	9, 9	10, 8
Puzzle 3	10, 2	8, 4
Puzzle 4	8, 7	9, 6
Puzzle 5	5, 5	9, 1
Puzzle 6	10, 6	12, 4

WORKING TOGETHER

- With a partner, discuss your strategies for solving *Circle Mysteries.*
- Find and record at least one more solution for each puzzle.
- Analyze your solutions. Describe, in writing and with pictures, any patterns you see.

CIRCLE MYSTERIES, PART 1

GAME BOARD

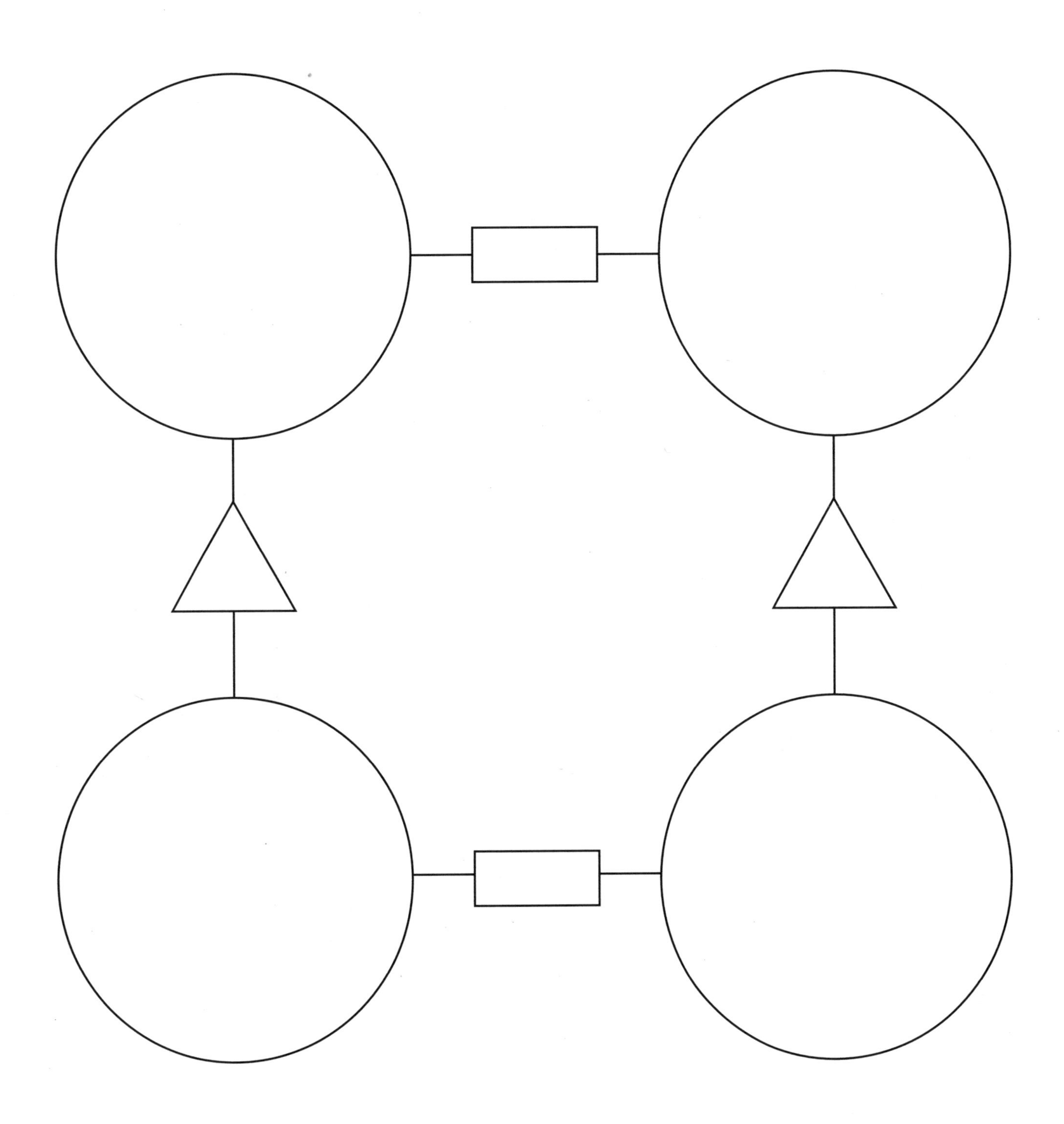

CIRCLE MYSTERIES, PART II

YOU NEED

- 20 counters
- Game board
- Recording sheets, page 69

HOW TO PLAY

Position counters in the circles on the game board so that the number in each rectangle is the same as the total number of counters in the two adjacent circles. For each puzzle:

- Fill in the rectangles with the given numbers.
- Use the numbers to figure out how many counters to put in each circle.
- If each given number is the sum of the counters in the adjoining two circles, you have found a solution.
- Record your solution.

Here is a sample puzzle and its solution.

Sample

Put these numbers in the rectangles: 5, 4, 7.

The solution requires 8 counters. Notice that the sum of the counters in Circles 1 and 2 is 7, the number in Rectangle C; the sum of the counters in Circles 2 and 3 is 4, the number in Rectangle B; and, the sum of the counters in Circles 1 and 3 is 5, the number in Rectangle A.

Solution

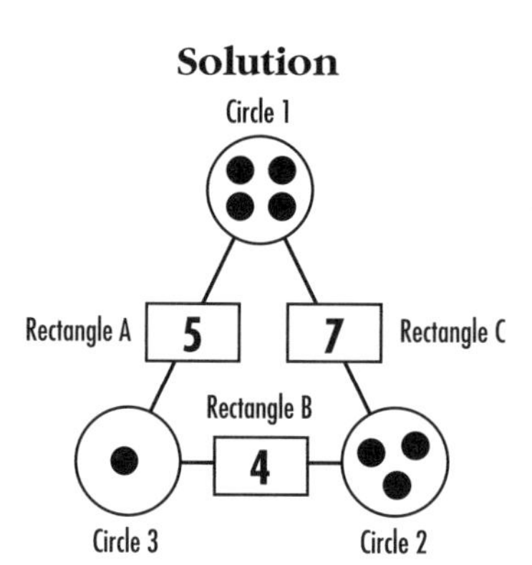

	Numbers for the rectangles		
Puzzle 1	6	6	8
Puzzle 2	5	6	9
Puzzle 3	8	9	11
Puzzle 4	6	9	15
Puzzle 5	10	15	17
Puzzle 6	7	8	11

WORKING TOGETHER

- With a partner, discuss strategies for solving *Circle Mysteries.*
- Analyze your solutions. Describe, in writing and with pictures, any patterns you see.

CIRCLE MYSTERIES, PART II

GAME BOARD

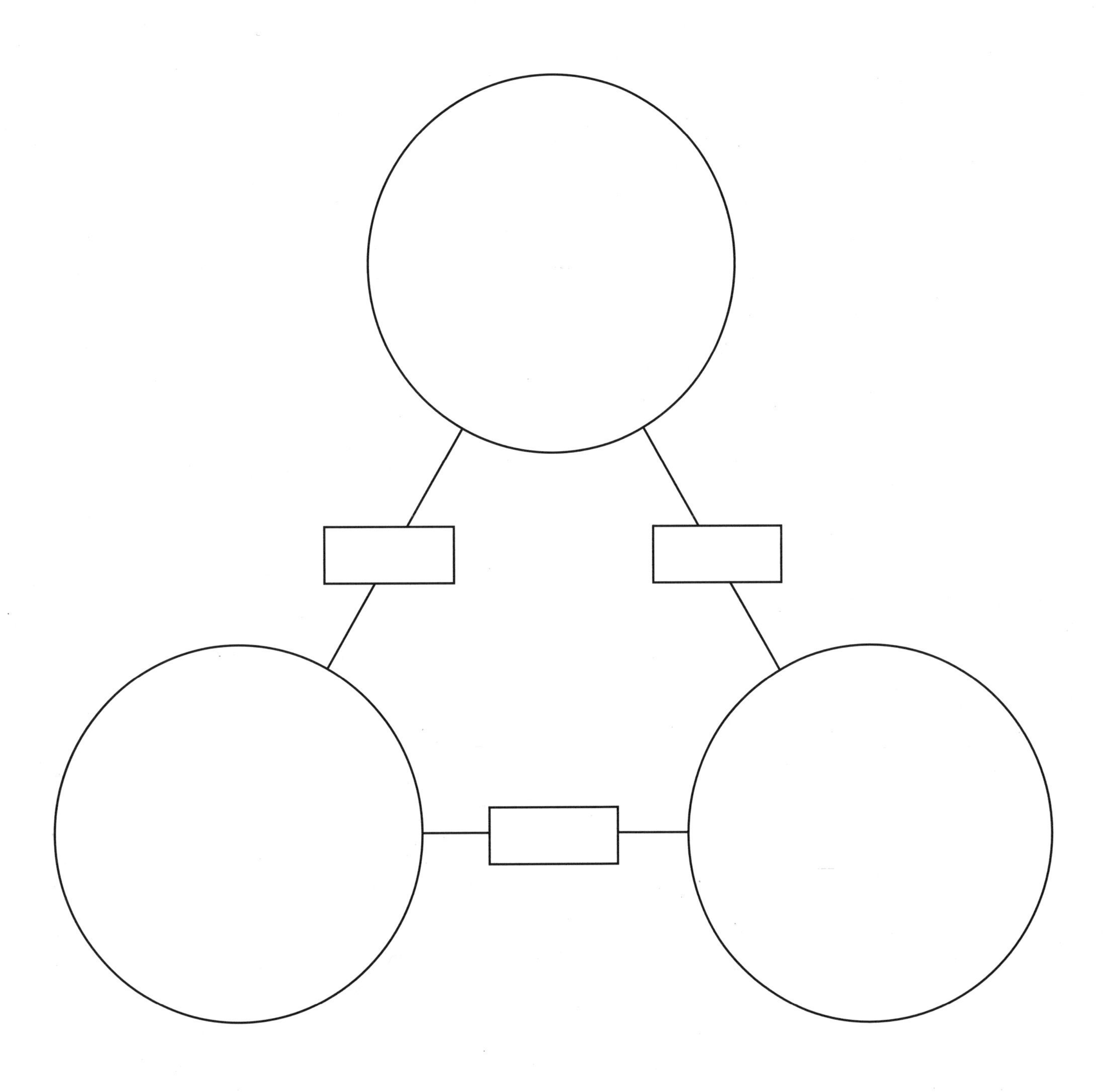

COORDINATED COLORS

YOU NEED

- 25 counters (5 each of 5 different colors)
- Game board

HOW TO PLAY

Place each of the 25 counters on the game board so that no color appears twice in any row, column, or diagonal.

WORKING TOGETHER

- Work with a partner to find two other solutions. Record your solutions on grid paper using colored pencils or crayons or invent a coding system of your own.
- Discuss, then write about, your strategies for solving Coordinated Colors.
- Try the same puzzle on a 7 x 7 game board (page 11) using 49 counters (7 each of 7 different colors). Find and record at least three solutions.

COORDINATED COLORS

GAME BOARD

SQUARE COUNT

YOU NEED

- 4 counters
- Game board

HOW TO PLAY

Determine the number of different ways 4 counters can be arranged on the game board so they form the corners of a square. Here are some examples.

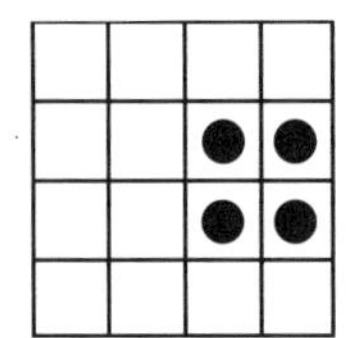

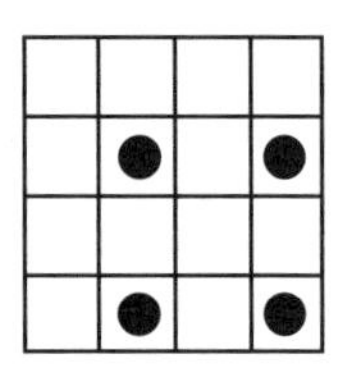

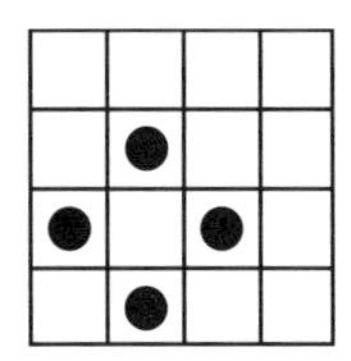

Now find out how many squares you can make altogether. Record your solutions.

Puzzle 1 Use the 4 x 4 game board.

Puzzle 2 Use the 5 x 5 game board.

WORKING TOGETHER

- With a partner, discuss this: How can you be sure that you have found all the solutions?
- Compare your solutions. Did you record your solutions in the same way?
- Describe, in writing and with pictures, strategies you used and patterns you noticed.

SQUARE COUNT

GAME BOARDS

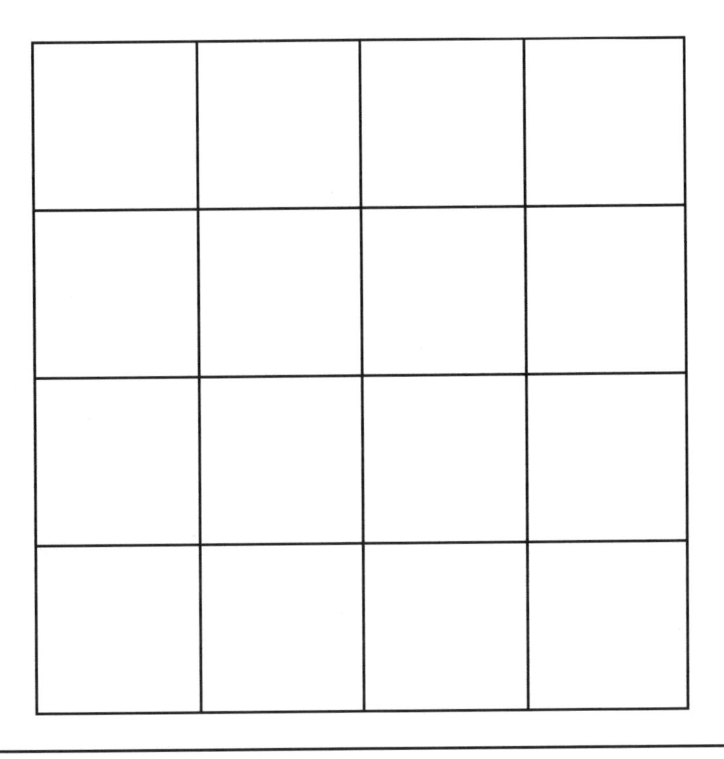

RING TEASERS

YOU NEED

- 15 counters
- Game board
- Recording sheet, page 70

HOW TO PLAY

Place the given number of counters on the game board. Position the counters in the sections of the rings so that the number of counters within each ring is the same as the number in the square in that ring. For each puzzle:

- Fill in the squares with the given numbers.
- Move the counters into different sections of the rings until you have a solution.
- Record your solution.
- Then, find another solution. Record that, too.

Here is a sample puzzle and two possible solutions.

Sample

Put these numbers in the squares: 3, 4, 5.

Notice that some counters are in more than one circle. For example, in Solution 1, Counter A is in two circles and Counter B is in three circles, whereas Counter C is in only one circle.

Solution 1

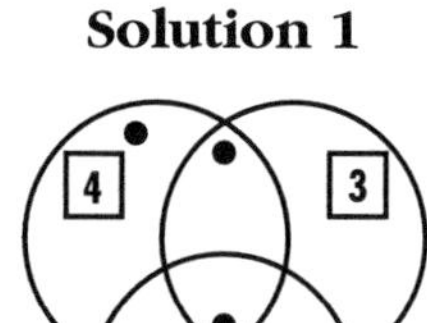

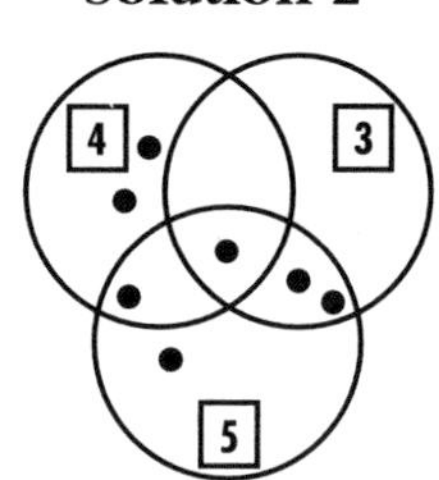

	Use this many counters	Put these numbers in the rings		
Puzzle 1	8	2	3	4
Puzzle 2	10	3	4	6
Puzzle 3	15	4	6	12
Puzzle 4	11	5	6	7
Puzzle 5	13	6	6	6

WORKING TOGETHER

- Working with a partner, imagine that the number of counters to use is not given. For each puzzle, find a solution that uses the fewest possible number of counters and one that uses the largest possible number. Record your solutions.
- Pick any puzzle above. Find all the possible solutions using various numbers of counters. Make a chart and/or draw pictures to show all your solutions. How do you know you have found all the solutions?

RING TEASERS

GAME BOARD

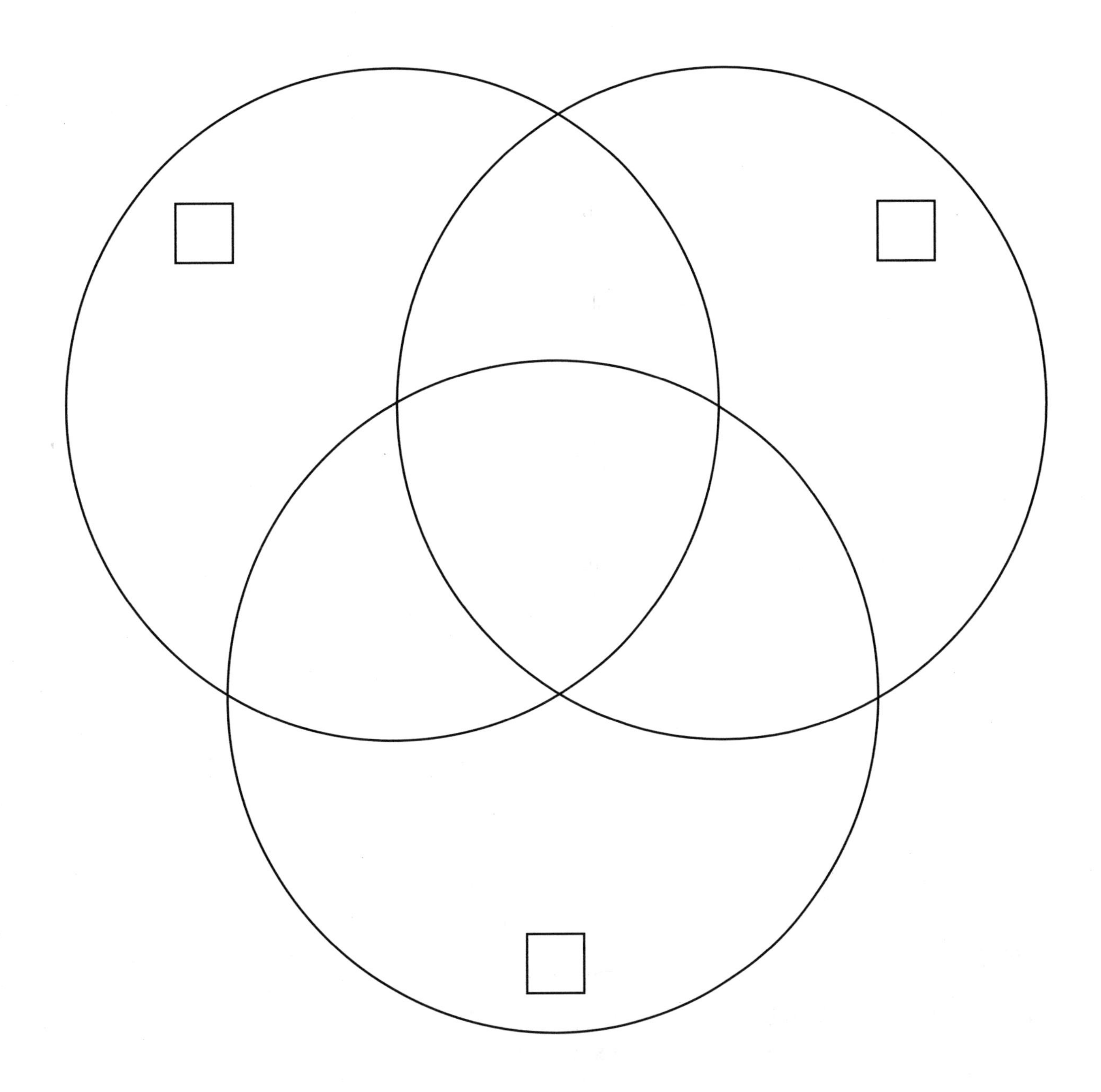

ODD EVEN SQUARES

YOU NEED

- 8 counters
- Game board
- Recording sheet, page 71

HOW TO PLAY

Put counters inside the squares on the game board so that the specified number of counters is in each row and column.

Here is a sample puzzle for a 5 x 5 grid and two possible solutions.

Sample

Position 5 counters so that each column and row contains an odd number of counters.

Solution 1

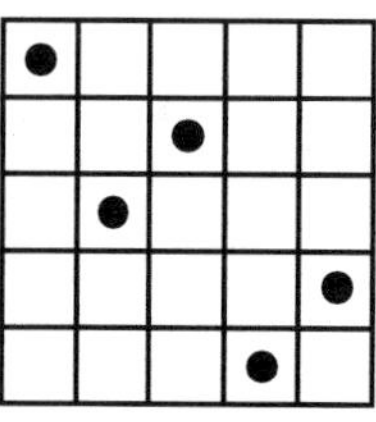

Solution 2

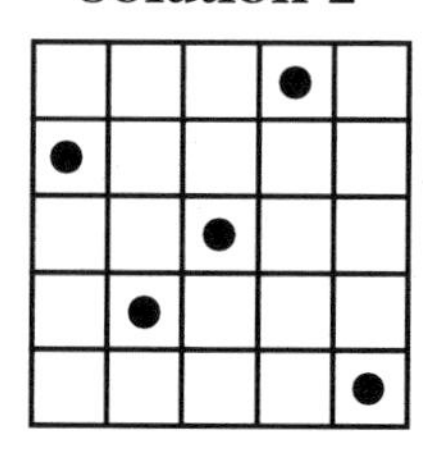

Use the 3 x 3 game board. Find and record at least two solutions for each puzzle.

Puzzle 1 Position 3 counters so that each column and each row contains an odd number of counters.

Puzzle 2 Position 5 counters so that each column and each row contains an odd number.

Puzzle 3 Position 6 counters so that each column and each row contains an even number.

Use the 4 x 4 game board. Find and record at least two solutions for each puzzle.

Puzzle 4 Position 4 counters on the game board so that each column and each row contains an odd number.

Puzzle 5 Position 6 counters so that each column and each row contains an odd number.

Puzzle 6 Position 8 counters so that each column and each row contains an odd number.

Puzzle 7 Position 8 counters so that each column and each row contains an even number.

WORKING TOGETHER

- Discuss with a partner whether or not it is possible to solve this problem: On a 3 x 3 grid, position 4 counters so that each column and each row contains an even number of counters.

ODD EVEN SQUARES

GAME BOARDS

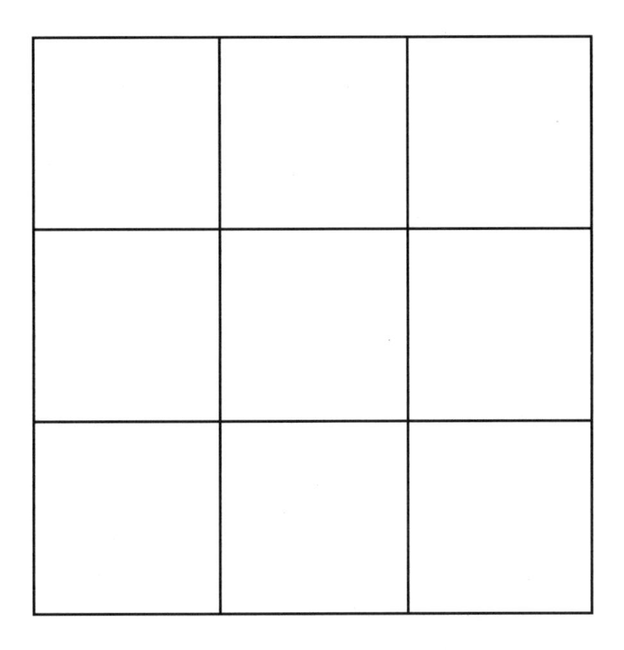

ODD EVEN CIRCLES

YOU NEED

- 6 counters
- Game board
- Recording sheet, page 72

HOW TO PLAY

Put counters inside the circles on the game board so that the specified numbers of counters is in each circle.

Here is a sample puzzle and two possible solutions.

Sample

Use 3 counters. Put an even number of counters in Circle A. Put an odd number of counters in Circle B.

Solution 1

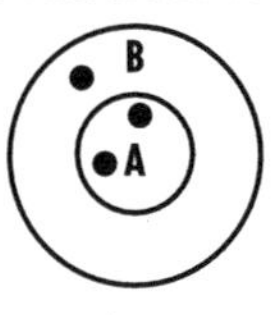

Solution 2

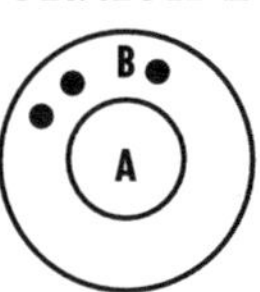

In solution 1, note that the 2 counters inside Circle A are also contained in Circle B, so Circle B has a total of 3 counters inside it. In solution 2, Circle A contains an even number of counters because 0 is an even number.

Use 6 counters. Find and record at least two solutions for each puzzle.

Puzzle 1 Put an even number in A.
Put an even number in B.
Put an even number in C.

Puzzle 2 Put an odd number in A.
Put an even number in B.
Put an even number in C.

Puzzle 3 Put an odd number in A.
Put an odd number in B.
Put an even number in C.

Puzzle 4 Put an even number in A.
Put an odd number in B.
Put an even number in C.

Use 5 counters. Find and record at least two solutions for each puzzle.

Puzzle 5 Put an odd number in A.
Put an odd number in B.
Put an odd number in C.

Puzzle 6 Put an even number in A.
Put an odd number in B.
Put an odd number in C.

Puzzle 7 Put an even number in A.
Put an even number in B.
Put an odd number in C.

Puzzle 8 Put an odd number in A.
Put an even number in B.
Put an odd number in C.

WORKING TOGETHER

- Discuss with a partner whether or not it is possible to solve this problem:
  Use 7 counters. Put an odd number in A. Put an odd number in B. Put an even number in C.

ODD EVEN CIRCLES

GAME BOARD

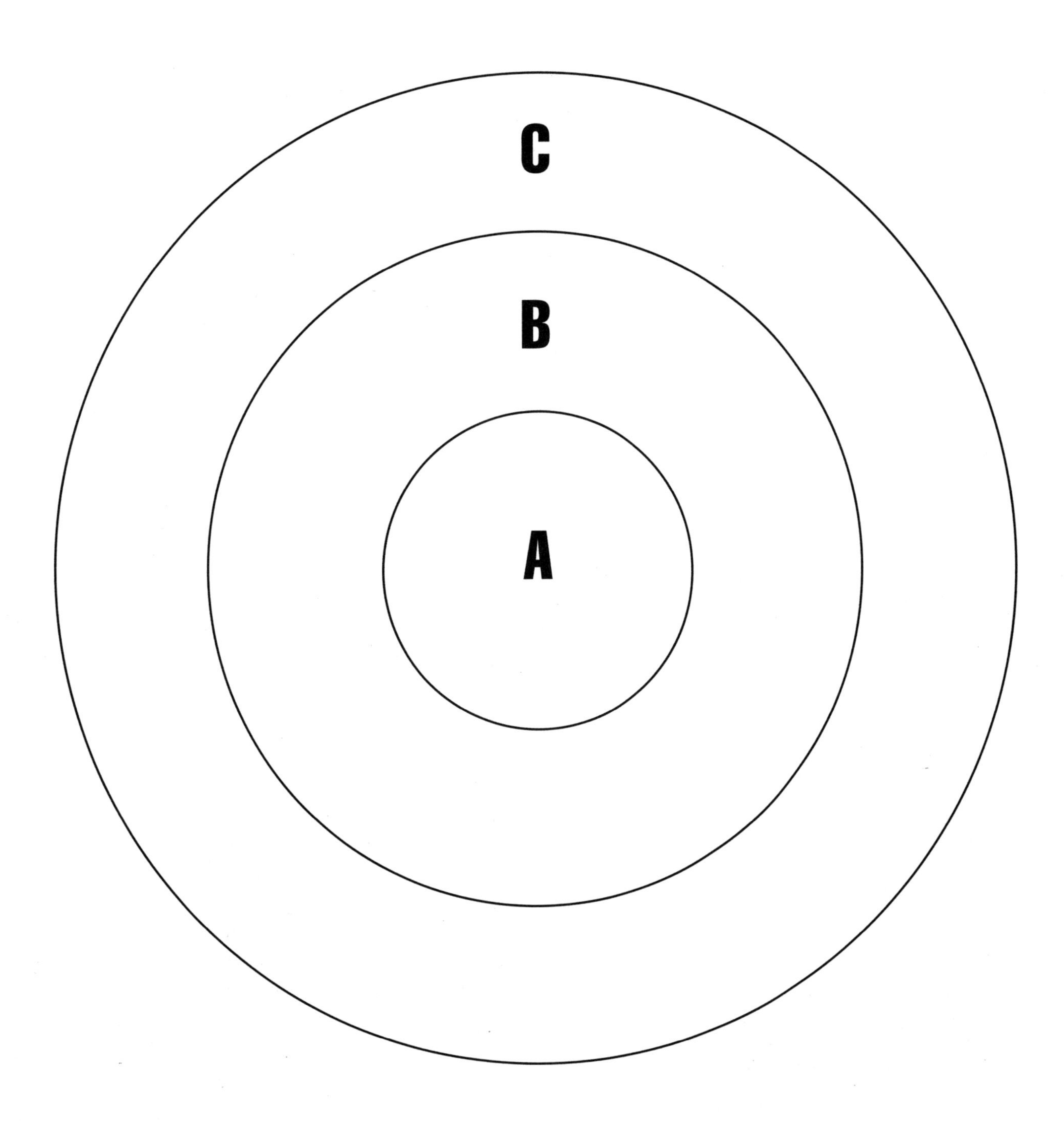

ODD EVEN SHAPES

YOU NEED

- 8 counters
- Game board
- Recording sheet, page 73

HOW TO PLAY

Put counters inside the shapes on the game board so that the specified number of counters is in each shape.

Here is a sample puzzle and its solution.

Sample

Use 5 counters.
Put an odd number in the circle.
Put an even number in the octagon.
Put an even number in the rectangle.

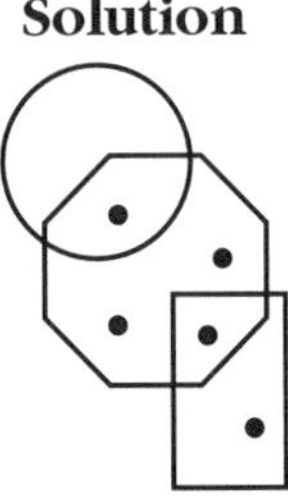

Notice that in the solution, the counter inside the circle and one of the counters inside the rectangle are also inside the octagon.

Use 3 counters for these puzzles. Record your answers on the recording sheet.

Puzzle 1
Put an even number in the circle.
Put an even number in the octagon.
Put an even number in the rectangle.

Puzzle 2
Put an even number in the circle.
Put an even number in the octagon.
Put an odd number in the rectangle.

Puzzle 3
Put an even number in the circle.
Put an odd number in the octagon.
Put an odd number in the rectangle.

Puzzle 4
Put an even number in the circle.
Put an odd number in the octagon.
Put an even number in the rectangle.

Puzzle 5
Put an odd number in the circle.
Put an odd number in the octagon.
Put an odd number in the rectangle.

Puzzle 6
Put an odd number in the circle.
Put an odd number in the octagon.
Put an even number in the rectangle.

Puzzle 7
Put an odd number in the circle.
Put an even number in the octagon.
Put an even number in the rectangle.

Puzzle 8
Put an odd number in the circle.
Put an even number in the octagon.
Put an odd number in the rectangle.

WORKING TOGETHER

- With a partner, use 7 counters to solve Puzzles 1 through 8. Then, try these puzzles using other numbers of counters from 1 to 10 (not 7 or 3).

ODD EVEN SHAPES

GAME BOARD

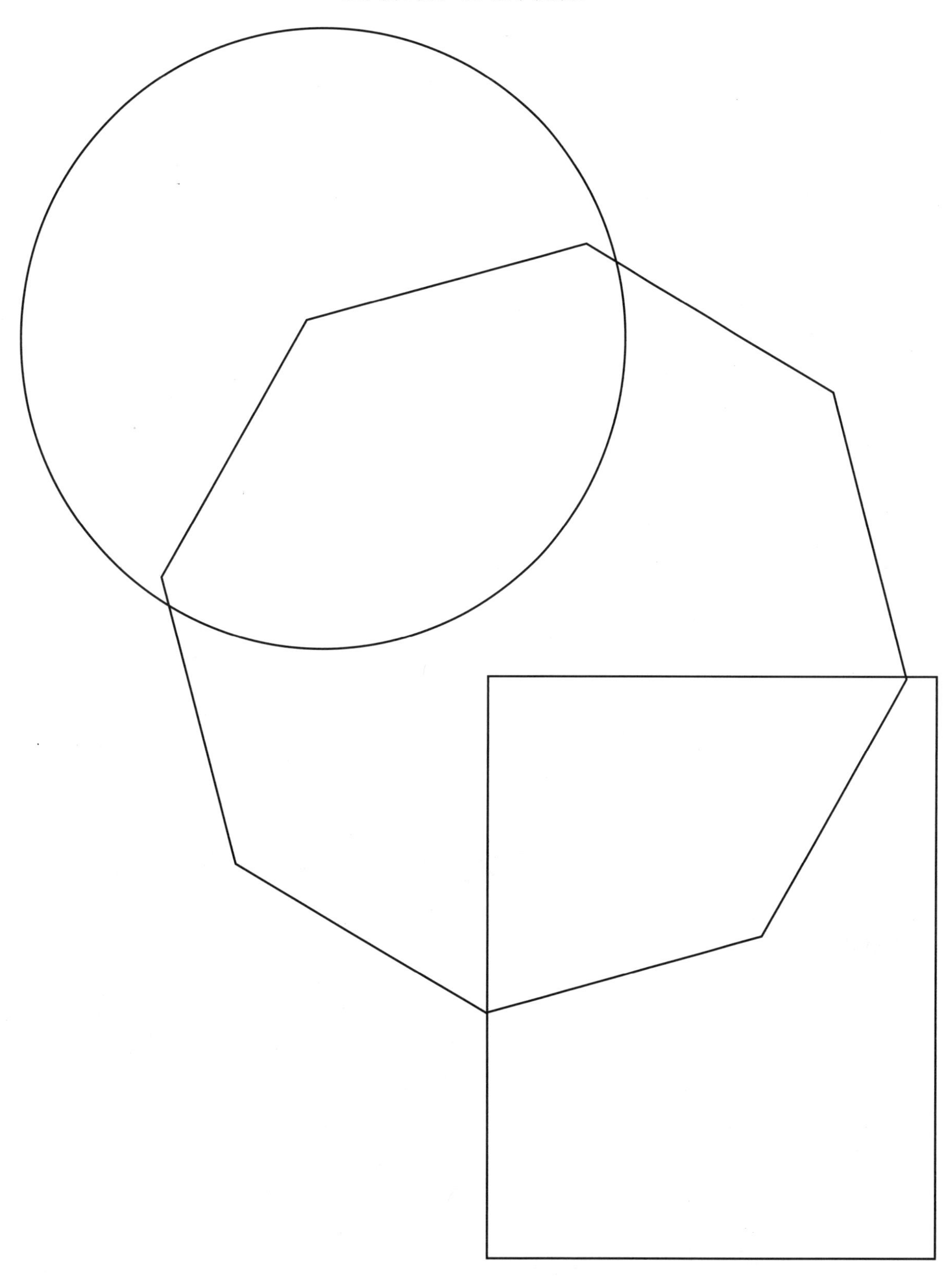

SWITCHING PLACES

YOU NEED

- 5 red counters
- 5 blue counters
- Game board

HOW TO PLAY

Pick a game board. Place an equal number of counters on each end of the board. On one end place red counters, and on the other end place blue counters. Leave the middle space empty.

Reverse the positions of the red counters and the blue counters using the minimum number of moves. Follow these rules:

- Only 1 counter can be moved at a time.
- A counter can move only into an empty space.
- A counter can jump over 1 counter of a different color. It cannot jump over its own color, nor can it jump over 2 counters.
- A counter that is not jumping can move only into an adjacent empty space.
- Record the number of moves necessary for the counters to switch places.

Play on each of the remaining game boards.

Here is a sample puzzle and its solution.

Sample

Use a 3-circle game board. Use 2 counters, 1 blue and 1 red. Leave the middle circle empty. Reverse the position of the blue counter and the red counter.

Solution

1ST MOVE 2ND MOVE 3RD MOVE

RED BLUE

BLUE RED

WORKING TOGETHER

- With a partner, make a chart to show the minimum number of moves needed to switch places on each of the different-sized game boards.
- Look for patterns that you can use to predict the minimum number of moves needed to switch 6 red counters and 6 blue counters; to switch 10 red counters and 10 blue counters.
- Determine the least number of moves necessary to switch any number of red counters and the same number of blue counters.

SWITCHING PLACES

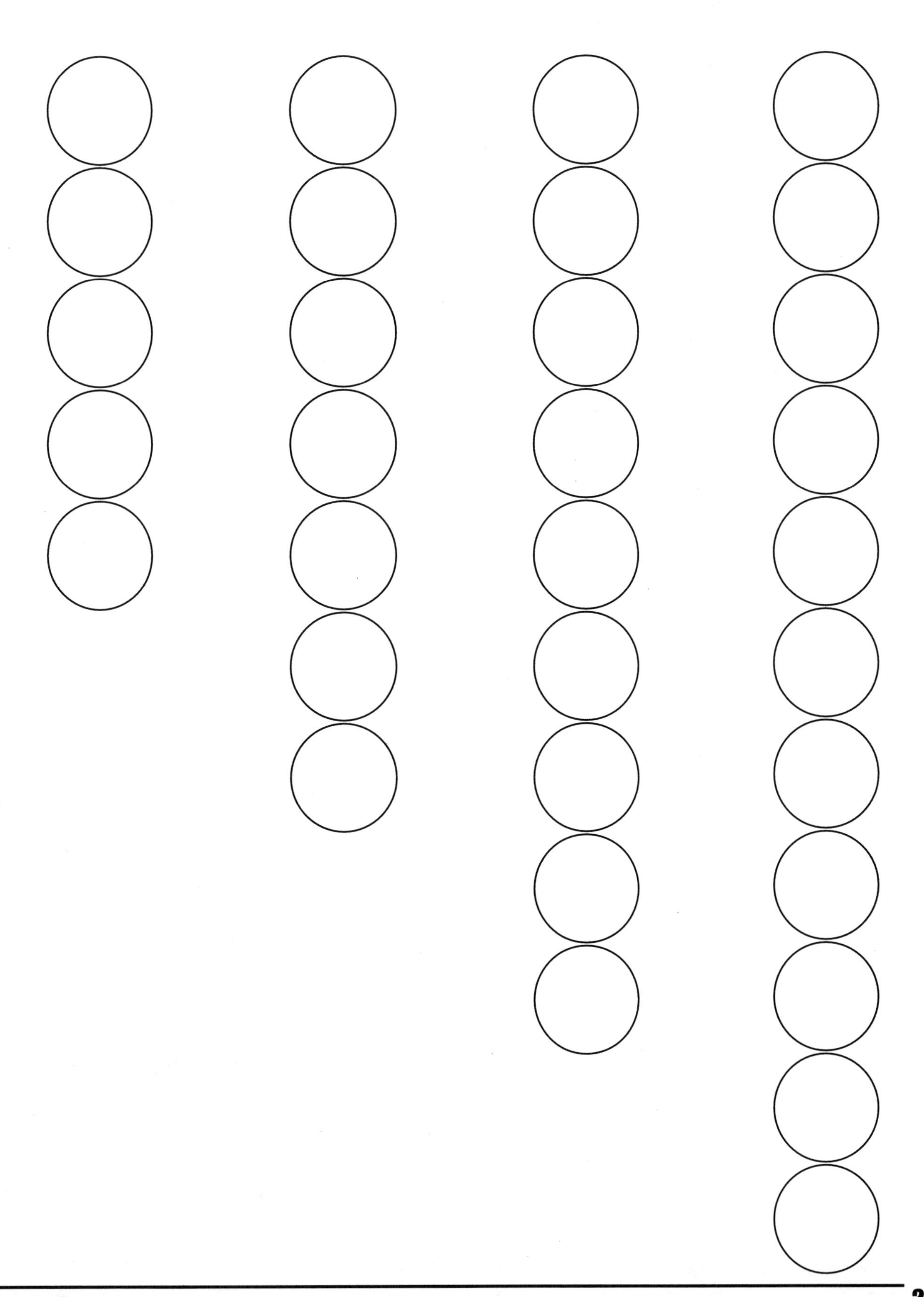

CRISS-CROSS

YOU NEED

- 7 counters
- Game board

HOW TO PLAY

Position 7 counters on the game board, according to these rules:

- Place a counter on any empty circle. A counter can only be placed on an empty circle.
- Slide the counter along one of the two possible pathways to an empty circle where it must remain.

Repeat with each of the remaining counters.

WORKING TOGETHER

- With a partner, discuss, then write about, strategies for solving Criss-Cross.

CRISS-CROSS

GAME BOARD

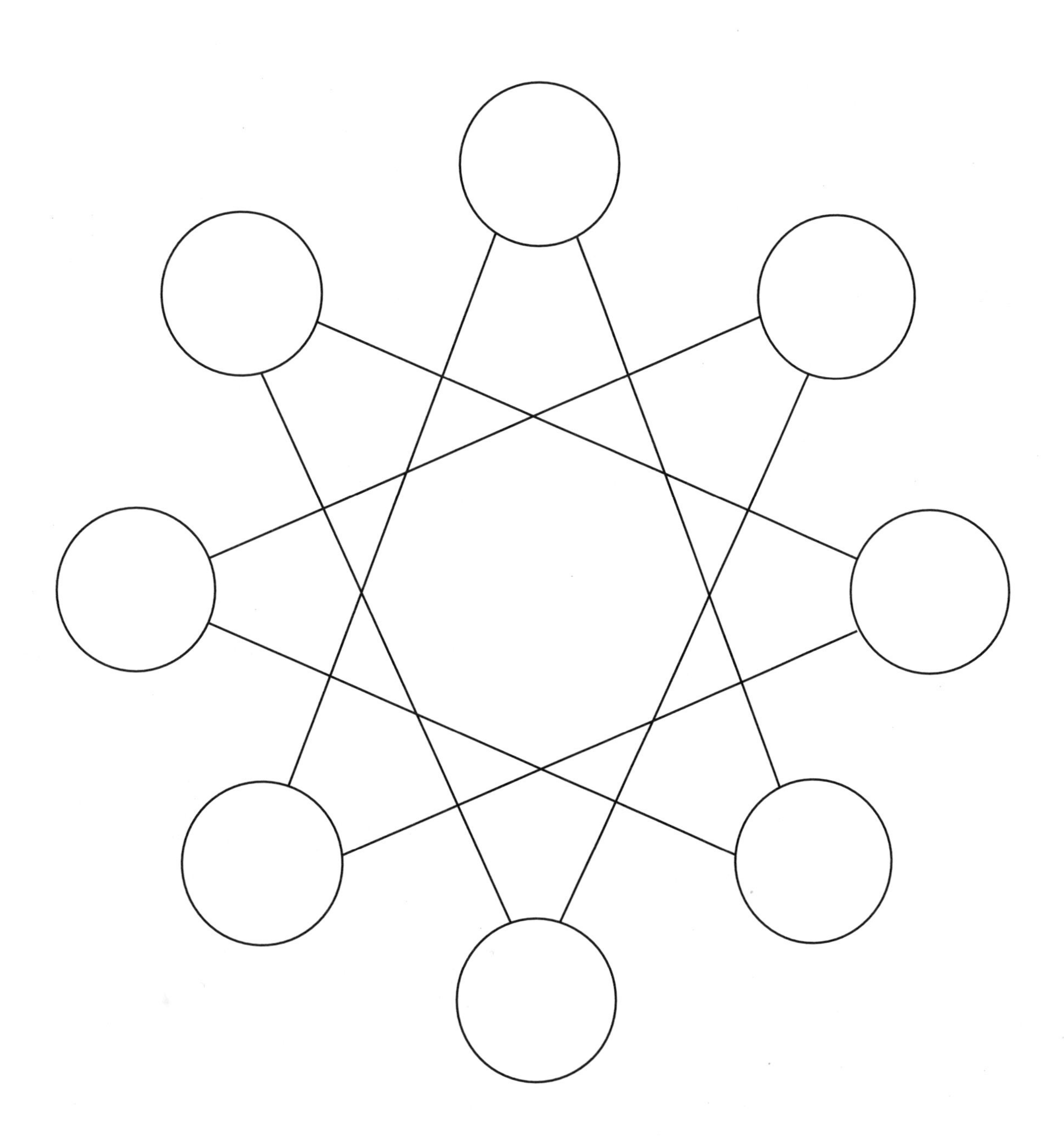

CRAZY CUPS

YOU NEED

- 11 counters
- 3 clear plastic cups

HOW TO PLAY

Place counters into the 3 cups so that each cup contains an odd number of counters.

Puzzle 1 Use 11 counters. Find as many solutions as you can find.

Puzzle 2 Use 10 counters. Find as many solutions you can.

WORKING TOGETHER

- With a partner, prepare a chart or picture to show all the solutions you found for each puzzle.
- Discuss this: Do you think that you have found all possible solutions? Why?

STUDENT ACTIVITIES

STRATEGY GAMES

BAD APPLE

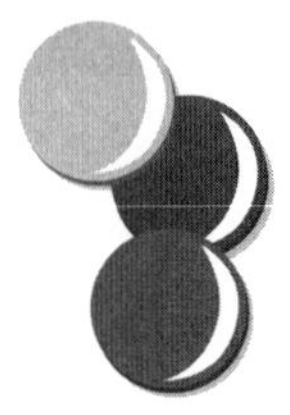

YOU NEED

- 15 counters of one color
- 1 counter of a different color (the Bad Apple)
- A partner

HOW TO PLAY

The object of the game is to force your opponent to take the Bad Apple (the last counter). Here are the rules:

- Place 12 counters of the same color on the game board, putting one in each "whole" apple. Place the counter of the different color in the "eaten" apple.
- Decide who goes first. Take turns.
- On your turn, remove 1, 2, or 3 counters.

The winner is the player who does not take the last counter.

Play the game at least five times.

WORKING TOGETHER

- Discuss strategies for winning Bad Apple.
- Decide if it is better to go first or second in this game. Explain in writing.
- Add 3 more apples to the game board. Play the game again, but this time use 15 counters—14 of one color and 1 of a different color. Do your strategies for winning the 130-counter game still work? Explain.

BAD APPLE

GAME BOARD

CIRCLE GAME

YOU NEED

- 10 counters
- Game board
- A partner

HOW TO PLAY

The object of the game is to take the last counter. Here are the rules:

- Put 1 counter on each of the 10 circles on the game board.
- Decide who goes first. Take turns.
- On your turn, remove 1 or 2 counters. If you remove 2 counters, they must be next to each other.

The winner is the player who takes the last counter.

Play the game at least five times.

WORKING TOGETHER

- Discuss strategies for winning the Circle Game.
- Draw 3 more circles on your game board. Play the game again, but this time use 13 counters. Do your strategies for winning the 10-counter game still work for this one. Explain.

CIRCLE GAME

GAME BOARD

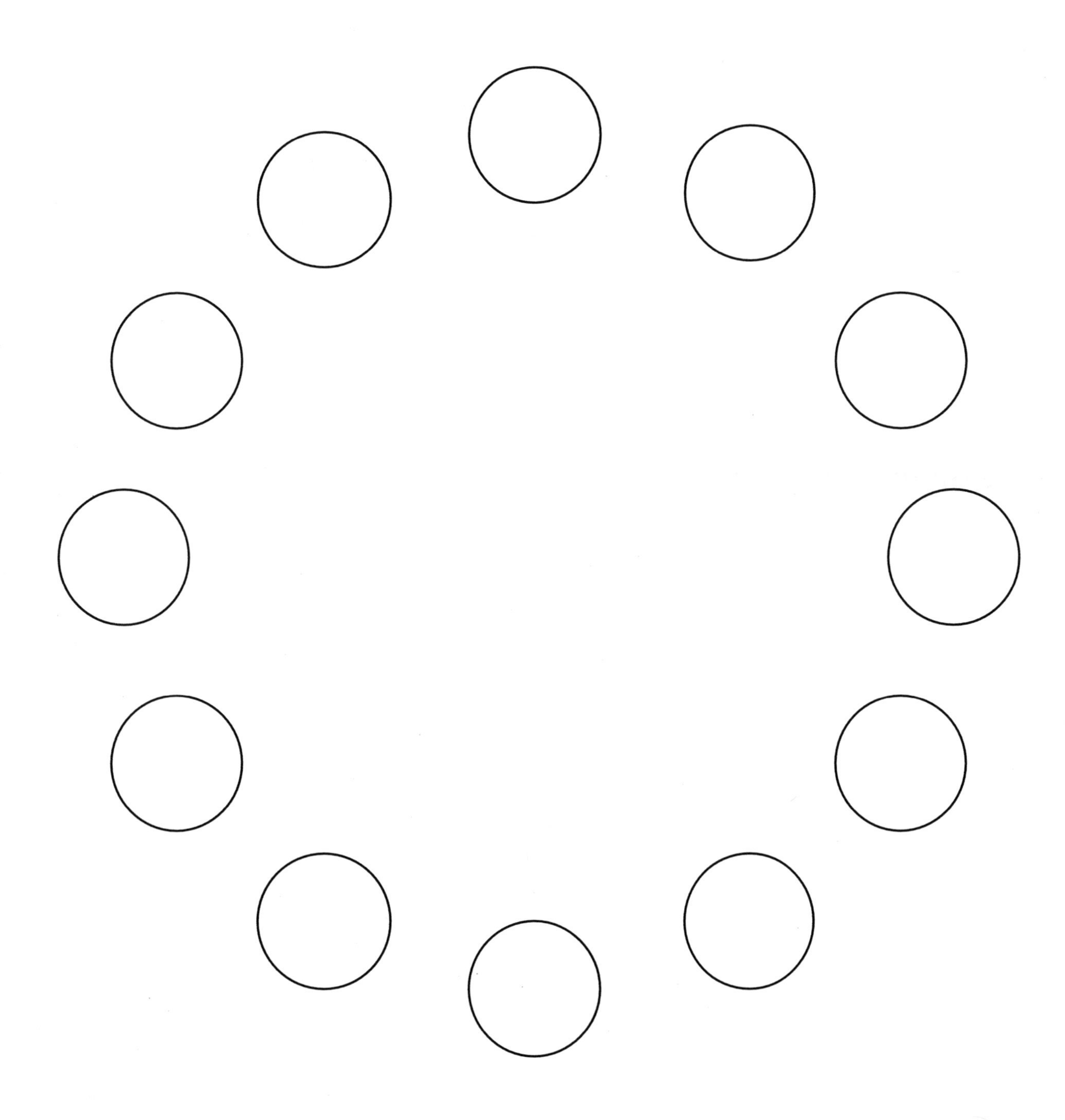

NIM

YOU NEED

- 12 counters
- Game board
- A partner

HOW TO PLAY

The object of the game is to take the last counter. Here are the rules:

- Set up 12 counters in three rows. Put 3 counters in the first row, 4 counters in the second row, and 5 counters in the last row.
- Decide who goes first. Take turns.
- On your turn, remove any number of counters as long as the counters come from the same row.

The winner is the player who takes the last counter.

Play the game at least five times.

WORKING TOGETHER

- Discuss strategies for winning Nim.
- Decide if it is better to go first or second. Explain in writing.
- Talk about which moves are good moves. Be prepared to describe why they are good.
- Determine at what point (or points) in the game it is possible to predict who is going to win.

NIM

GAME BOARD

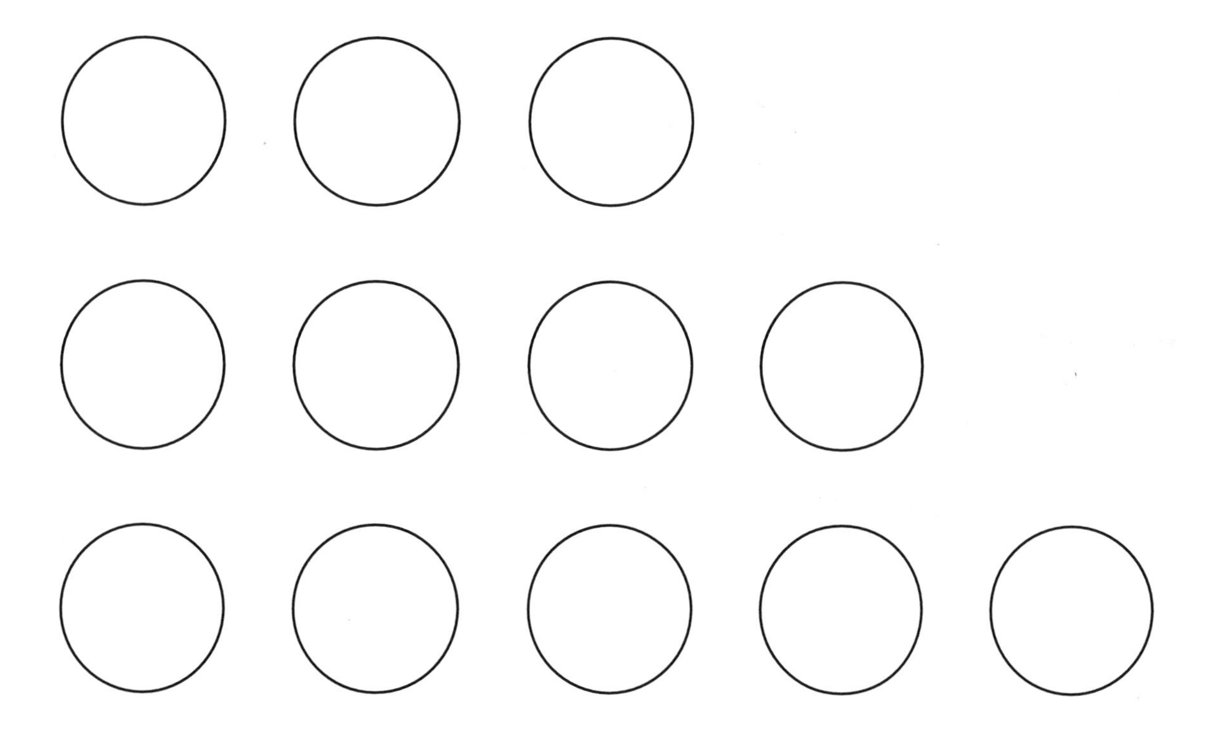

LINE UP

YOU NEED

- 40 counters (each player needs 20 of a different color)
- Game board
- A partner

HOW TO PLAY

The object of the game is to place 3 counters adjacent to each other, either vertically, horizontally, or diagonally. Here are the rules:

- Decide who goes first. Take turns.
- On your turn, place a counter on any empty square.
- Once a counter has been placed, it cannot be moved.

The winner is the first player to place 3 counters of the same color next to each other in a row, in a column, or on the diagonal.

Play the game at least five times.

WORKING TOGETHER

- Discuss strategies for winning Line Up.
- Play the game again, but this time use 4 counters. Do your strategies for winning the 3-counter game still work? Explain.

LINE UP

GAME BOARD

HEX

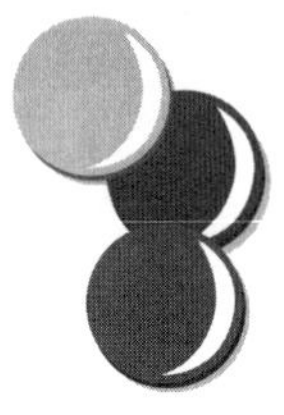

YOU NEED

- 16 counters (each player needs 8 of a different color)
- Game board
- A partner

HOW TO PLAY

The object of the game is to construct an unbroken path from one side of the game board to the opposite side. Here are the rules:

- Select a game board.
- Decide who goes first. Take turns.
- On your turn, place a counter on any empty hexagon.
- Once a counter has been placed, it cannot be moved.
- Corner hexagons count for either player.

The winner is the first player to construct an unbroken path from one side of the board to the other.

Play the game at least five times.

The sample game shown was played on a 3 x 3 game board. Player 1 won by creating a path of 3 black counters across the board.

Sample

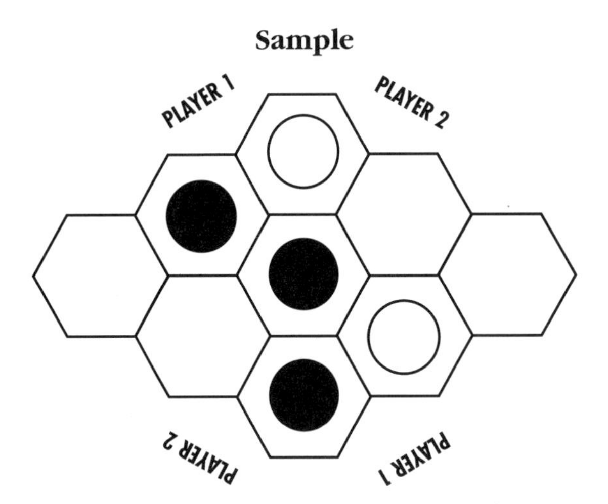

WORKING TOGETHER

- Discuss strategies for winning Hex.
- Decide if you would rather go first or second. Explain.
- For each game board, determine the fewest turns necessary to win.

HEX

GAME BOARDS

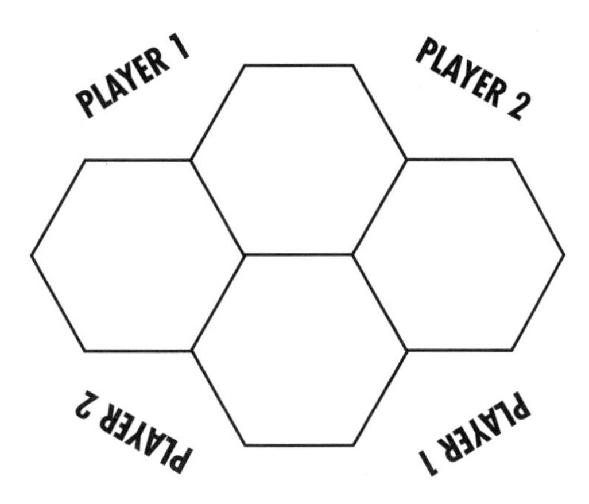

PLAYER 1
PLAYER 2
PLAYER 2
PLAYER 1

HEX

GAME BOARDS

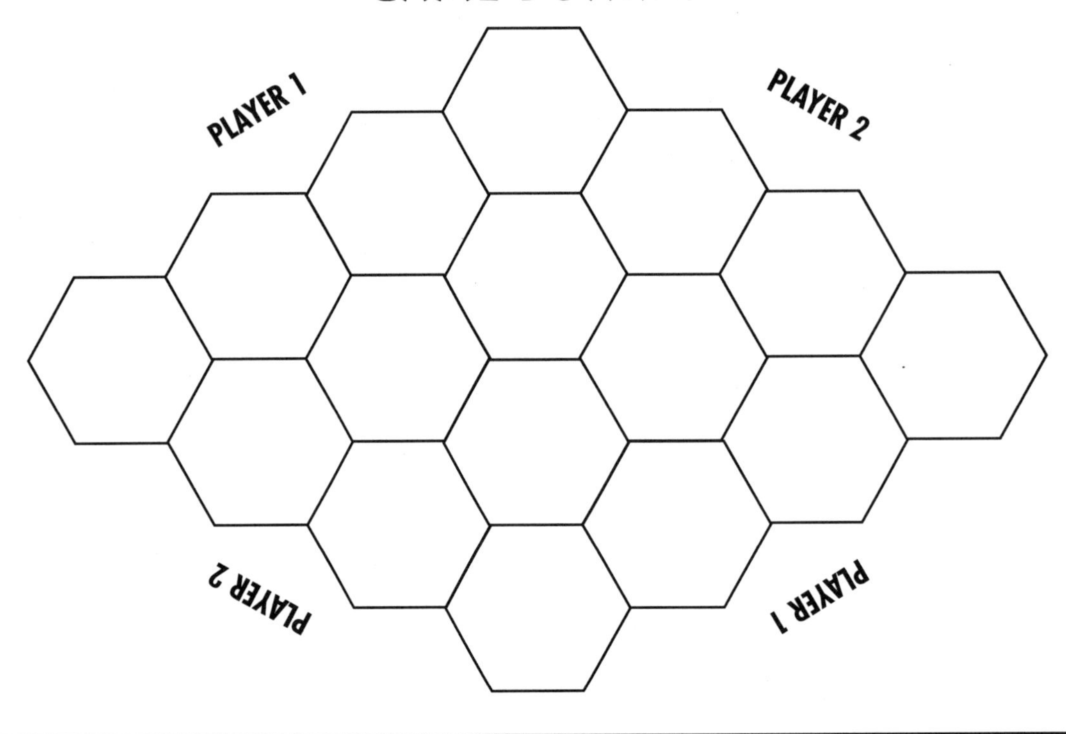

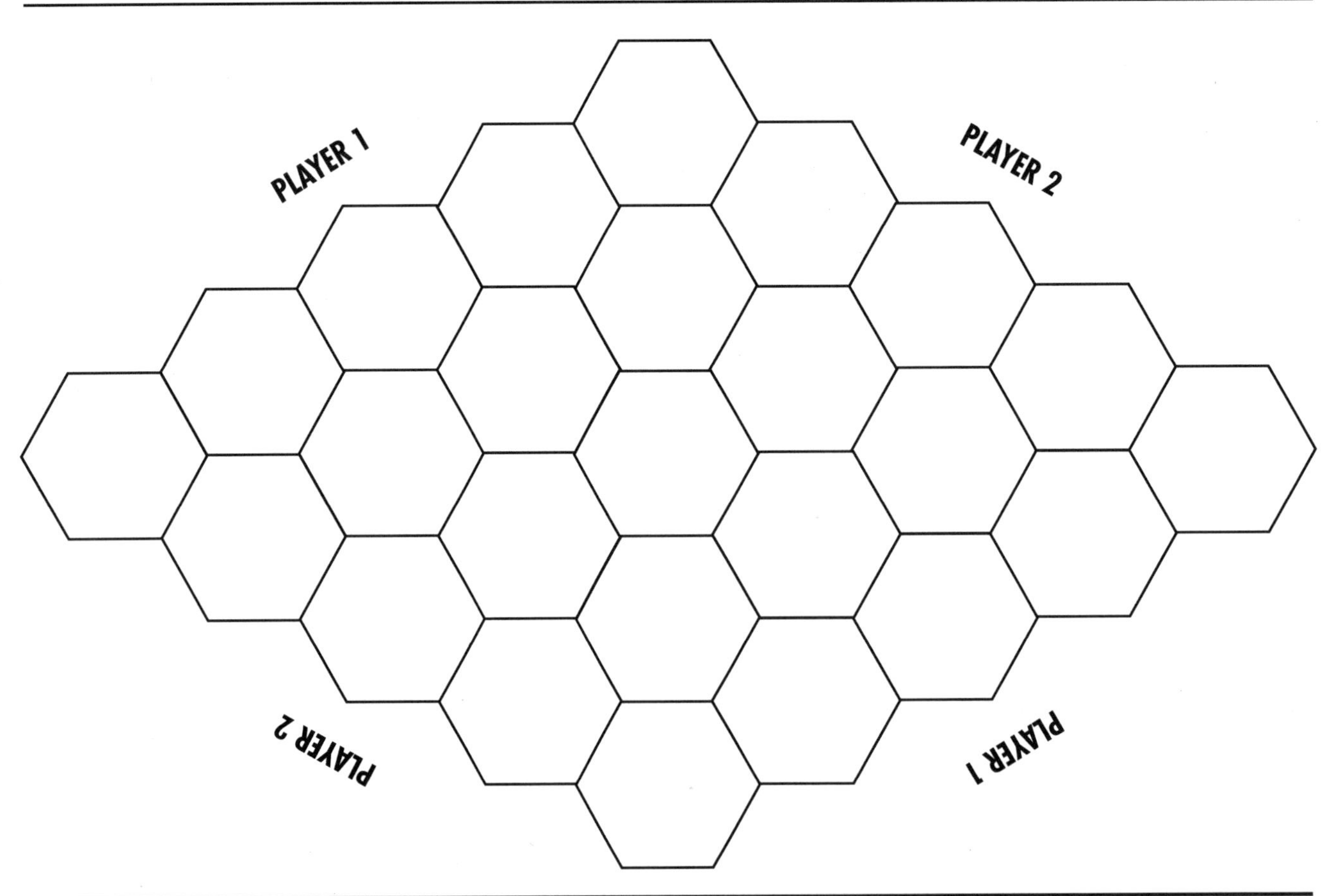

TEACHER NOTES

LEAVE ONE

Page 6

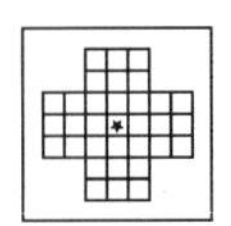

Leave One is a collection of solitaire games played on a grid-like board. The board has 33 squares on which the counters are placed. The object of Leave One is to remove all the counters except 1, which finishes in the center square.

Solutions

When you are recording moves, it is helpful to label the board. One way is to think of the board as a grid. As shown in the example, each square is given a label: the letter tells the horizontal row, and the number designates the vertical column.

Another way to record moves is to simply number each square from 1 to 33.

In the solutions given below, the first coordinate indicates the starting position of the counter, and the second coordinate indicates where it lands (using the grid-labeling system explained above). For example, C4–C2 means the counter on square C4 jumped to square C2. This would remove the counter on square C3.

	1	2	3	4	5	6	7
A			A3	A4	A5		
B			B3	B4	B5		
C	C1	C2	C3	C4	C5	C6	C7
D	D1	D2	D3	D4	D5	D6	D7
E	E1	E2	E3	E4	E5	E6	E7
F			F3	F4	F5		
G			G3	G4	G5		

Puzzle 1

Jumps: 5
Jumps: C4–C2, E4–C4, C5–C3, C2–C4, B4–D4

Puzzle 2

Jumps: 8
Jumps: E4–G4, C4–E4, 64–D4, D4–F4, D2–D4, G4–E4, E4–C4, B4–D4

Puzzle 3

Jumps: 8
Jumps: E5–C5, C5–C3, E3–E5, E6–E4, D4–F4, C3–E3, E2–E4, F4–D4

Puzzle 4

Jumps: 11
Jumps: B5–D5, D4–D6, B3–D3, E1–E3, B4–D4, D3–F3, D4–F4, F3–F5, E7–E5, F5–D5, D6–D4

The last 2 counters in each game must end up in the positions shown here, or a rotation of it.

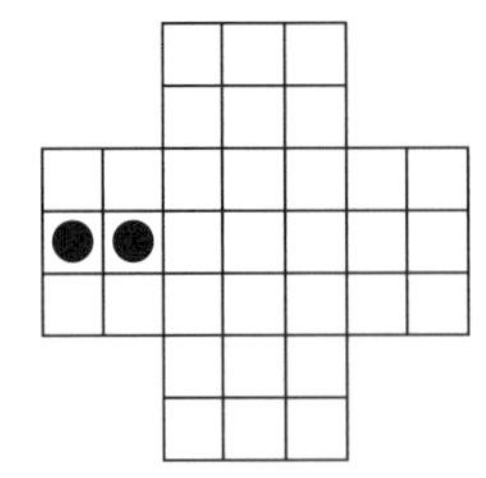

The last 3 counters need to be in one of these two positions, or a rotation or reflection of them.

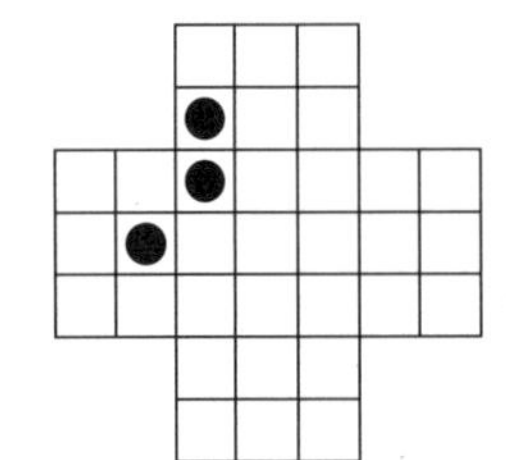

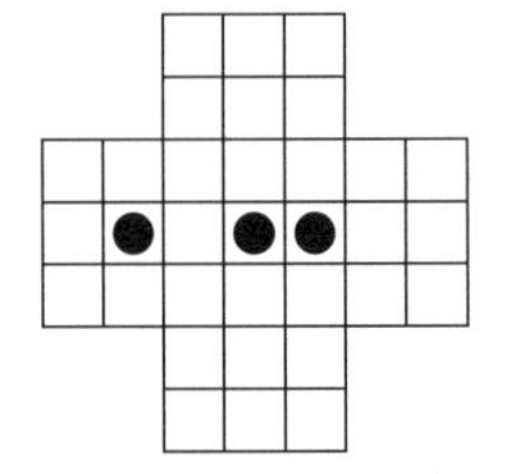

Variations

1. Students may enjoy creating their own Leave One puzzle. Many arrangements of counters are possible that leave 1 counter in the center square. (Students may find it easier to create a game by working backward.)
2. For a simpler version of the game, ask students to see how few counters they can leave on the board.

TIGHT FIT

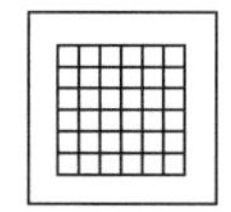

Page 8

Students are challenged to arrange counters on a grid so that no more than 1 counter (or 2 counters for the second activity) is in each column, row, or diagonal. To do so, they must double-check the placements of the counters as each new counter is added.

Solutions

Puzzle 1 The maximum number of counters is 6. One solution is shown, but others are possible.

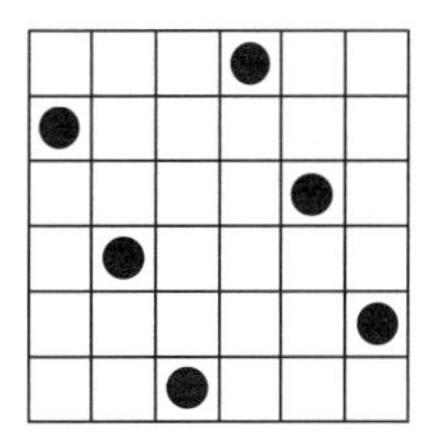

Puzzle 2 The maximum number of counters is 12. One solution is shown, but others are possible.

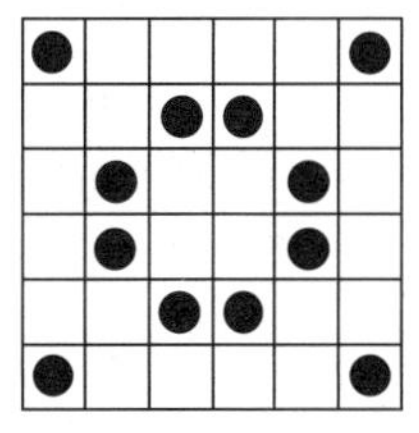

Strategy

Some students find it helpful to place a counter of a different color on the squares that cannot be used. This helps them to see which squares are available and which are not.

Patterns

As students try different-sized boards, they may realize that for Puzzle 1, the maximum number of counters is generally equal to the length of the side of the board. If a 4 x 4 board is used, then 4 counters is the maximum. However, when the board is 3 x 3 or 2 x 2, the solution is one less than the side length.

For Puzzle 2, on 2 x 2 boards and larger, the maximum number of counters is always equal to twice the length of the side.

Variations

1. For a simpler version of these puzzles, tell the students how many counters to use in each (6 in Puzzle 1, 12 in Puzzle 2).
2. Turn this into a two-player game, with both players using counters of the same color. The idea is to force your opponent to place a counter that makes a row, column, or diagonal containing 3 counters. Here are the rules:
   - Decide who goes first. Take turns.
   - On your turn, place a counter in any empty square on the game board.
   - Continue taking turns until one player places a counter that makes a row, column, or diagonal (not necessarily adjacent) containing 3 counters. The other player is the winner.

Have the students play at least five times and then answer these questions:

- What was the greatest number of counters that you and your partner placed on the game board before someone made a row of 3?
- What are some strategies for winning this game?

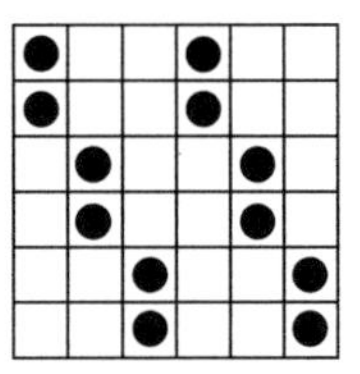

Variation Solution

The maximum number of counters that can be placed on the game board before a row of 3 is made is 12. One possible configuration is shown here.

Variation Strategy

Early in the game, students tend to focus more on the rows and columns, trying to find an empty space that will work. As the game proceeds, players again look for a row or column that has only 1 or 2 counters in it, and after finding a suitable row or column, they then look at the diagonals.

COUNTER MOVES

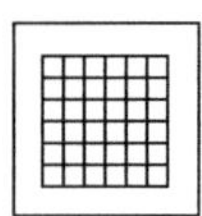

Page 10

Students are challenged to alter an arrangement of counters to fit specific requirements. To solve Puzzle 1, students must visualize each move and plan ahead as they transform the shape. For Puzzle 2, students must think about and visualize lines other than those formed along the rows, columns, and diagonals. The solution features lines with differing slopes.

Solutions

Puzzle 1 Five moves are necessary. Each counter in the picture is numbered to show the progression of moves. If you are allowed to jump diagonally, four moves are needed.

BEGIN

①			⑥
②	④	⑤	⑦
③			⑧

1ST MOVE

①		⑥	
②	④	⑤	⑦
③			⑧

2ND MOVE

①	④	⑥	
②		⑤	⑦
③			⑧

3RD MOVE

①	④	⑥	
②		⑤	⑦
③	⑧		

4TH MOVE

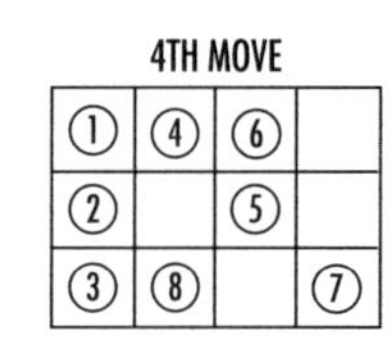

①	④	⑥	
②		⑤	
③	⑧		⑦

5TH MOVE

①	④	⑥	
②		⑤	
③	⑧	⑦	

Puzzle 2 Students usually can create a solution that has eight lines as shown here (the white counters show the 2 counters that were added). However, moving those 2 counters to make 10 lines is very challenging.

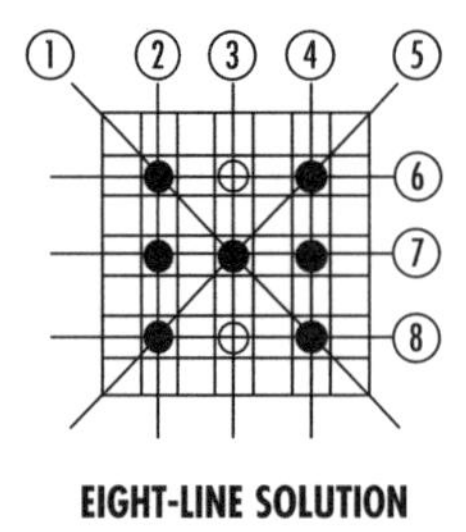

EIGHT-LINE SOLUTION

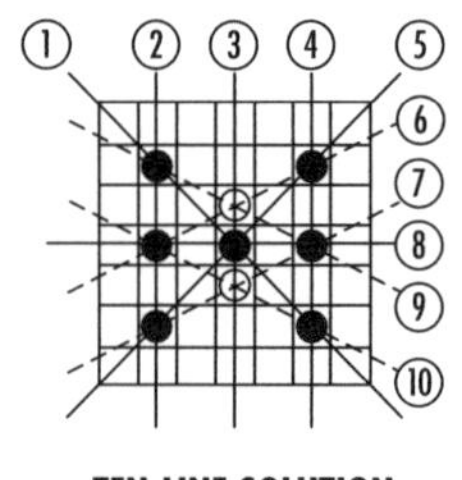

TEN-LINE SOLUTION

CIRCLE MYSTERIES, PARTS I AND II

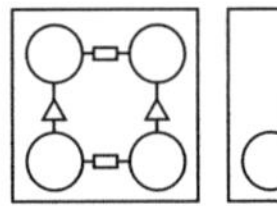
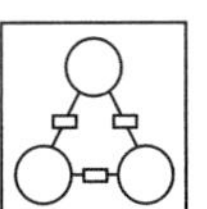

Pages 12,14

For the Circle Mysteries, Parts I and II, students try to discover the mystery number of counters placed in each circle. As they solve the puzzles, students may uncover many patterns and shortcuts for each type of Circle Mysteries. If students were using numbers rather than counters in these puzzles, it would be possible to solve some of the puzzles in Part I using fractions, decimals, or positive and negative numbers. For example, if students have been exploring operations on integers, you might ask them to create solutions using positive and negative numbers. Recording sheets are found on pages 68–69.

PART I

Solutions

Here are two solutions for each puzzle.

	PUZZLE 1	PUZZLE 2	PUZZLE 3	PUZZLE 4	PUZZLE 5	PUZZLE 6
Solution 1, top (circle–rectangle–circle)	5 – 12 – 7	4 – 9 – 5	7 – 10 – 3	8 – 8 – 0	4 – 5 – 1	3 – 6 – 1
Solution 1, triangles (left, right)	14, 7	8, 10	8, 4	9, 6	9, 1	4, 12
Solution 1, bottom (circle–rectangle–circle)	9 – 9 – 0	4 – 9 – 5	1 – 2 – 1	1 – 7 – 6	5 – 5 – 0	1 – 10 – 11
Solution 2, top (circle–rectangle–circle)	6 – 12 – 6	3 – 9 – 6	6 – 10 – 4	6 – 8 – 2	5 – 5 – 0	1 – 6 – 5
Solution 2, triangles (left, right)	14, 7	8, 10	8, 4	9, 6	9, 1	4, 12
Solution 2, bottom (circle–rectangle–circle)	8 – 9 – 1	5 – 9 – 4	2 – 2 – 0	3 – 7 – 4	4 – 5 – 1	3 – 10 – 7

Strategy

Students often use trial and error to solve the puzzles. As they try more puzzles, encourage students to look for patterns, such as the relationship between the number of counters and the numbers in the triangles and rectangles.

Patterns

Students may uncover a variety of patterns. Here are a few possibilities:

1. The sum of the four rectangle and triangle numbers in each puzzle is twice the number of counters needed in the circles. For example, in Puzzle 1, the total of the rectangle and triangle numbers is 42, and the total number of counters needed is 21. Since each circle is connected to both a rectangle and a triangle, each circle number is used in the total for two rectangle numbers and two triangle numbers. Therefore, the number in each circle is counted twice, making the sum of the rectangle and triangle numbers double the sum of the counters.
2. The total number of counters in both rectangles (or the total number of counters in both triangles) equals the total number of counters needed to solve the puzzle. This can be explained algebraically. In the diagram, R1 and R2 represent the numbers in Rectangles 1 and 2, T1 and T2 represent the numbers in Triangles 1 and 2, and A, B, C, D represent the number of counters in the circles.

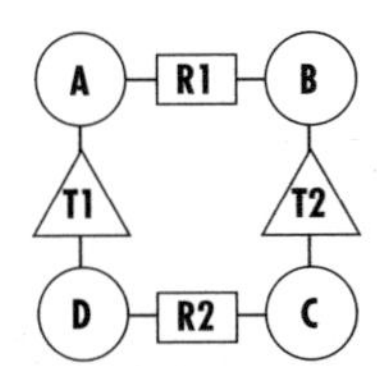

$$R1 = A + B$$
$$R2 = C + D$$

Therefore, $R1 + R2 = A + B + C + D$

3. The sum of the two rectangle numbers equals the sum of the two triangle numbers. This can be explained using the diagram on the bottom of page 49.

$$R1 = A + B$$
$$R2 = C + D$$
$$T1 = A + D$$
$$T2 = B + C$$
$$R1 + R2 = A + B + C + D$$
$$T1 + T2 = A + B + C + D$$
$$\text{Therefore, } R1 + R2 = T1 + T2.$$

Variations

1. Challenge students to solve each of the puzzles using either integers or fractions.
2. Have students create their own square-shaped Circle Mysteries for classmates to try. They must include at least one possible answer. Ask them to determine if any four numbers can be used to make the puzzles work. (Four numbers will work only if the sum of the two numbers in the triangles equals the sum of the two numbers in the rectangles.)

PART II

Solutions

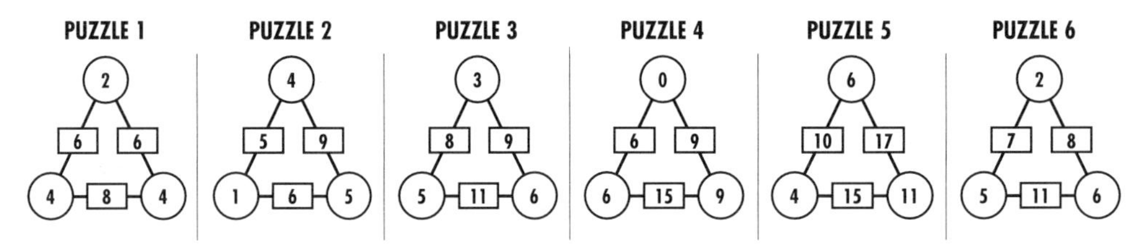

Strategy

Students often use trial and error to solve these puzzles, just as in Part I. As they try more puzzles, encourage them to look for patterns, such as the relationship between the number of counters and the numbers in the rectangles.

Patterns

Students may uncover a variety of patterns. Here are a few possibilities.

1. The sum of the three rectangle numbers in each puzzle is twice the number of counters needed in the circles. For example, in Puzzle 1, the total of the rectangle numbers is 20, and the total number of counters needed is 10. Since each circle is connected to two rectangles, each circle number is used for the total of two different rectangle numbers. Therefore, the number in each circle is counted twice, making the sum of the rectangle numbers double the sum of the counters.

2. The number of counters in a circle added to the number in the opposite rectangle will equal the total number of counters. For example, in Puzzle 1, the total number of counters to use is known from the first pattern: The sum of the rectangle numbers is 20, so the sum of the counters must be 10. Looking at the 8 in the rectangle and the circle opposite it, the total must be 10, so the circle must contain 2 counters. Likewise, the circles opposite each 6 must contain 4 counters. The sum of the circle numbers (4 + 4 + 2) equals 10.

Variations

1. Challenge students to make up their own triangular-shaped Circle Mysteries. Ask students if any three numbers can be placed in the rectangles to make the puzzles work. (Any three numbers will work. However, for some, integers or decimals and fractions must be used to solve the mystery.)
2. Make Circle Mysteries with other shapes, such as pentagons or hexagons.

COORDINATED COLORS

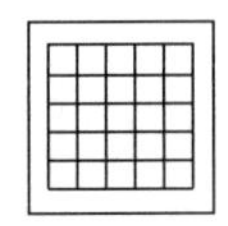

Page 16

Students are challenged to place 25 counters on the square game board so that no more than 1 counter of any color appears in any row, column, or diagonal. The puzzle may perplex students; often they will get close to the end and realize that they cannot correctly place the last few counters.

Solutions

More than one solution is possible. Here are several.

A	B	C	D	E
C	D	E	A	B
E	A	B	C	D
B	C	D	E	A
D	E	A	B	C

A	D	B	E	C
B	E	C	A	D
C	A	D	B	E
D	B	E	C	A
E	C	A	D	B

C	E	B	D	A
A	C	E	B	D
D	A	C	E	B
E	B	D	A	C
B	D	A	C	E

D	C	A	B	E
B	E	D	C	A
E	D	C	A	B
C	A	B	E	D
A	B	E	D	C

Strategy

One strategy is to place color A in the upper left column and then proceed by placing the next color A in a square over one and down two. Continue with this pattern to fill in all color A. Continue with the "over one, down two" pattern to fill in the other colors. This pattern also works for the 7 x 7 puzzle.

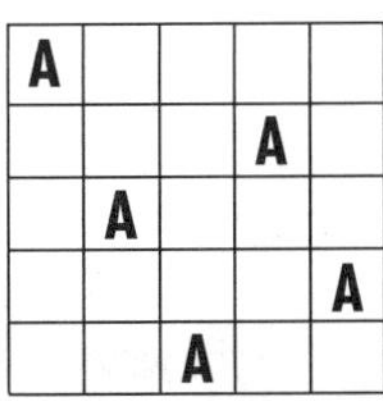

"OVER ONE, DOWN TWO"

Another strategy is to place a whole row of different colors (A-B-C-D-E) in the first row. For the next row, shift over two colors to begin with C (C-D-E-A-B). Continue with this pattern, shifting over two colors for each row.

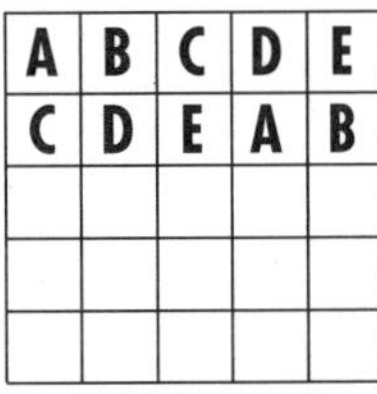

"WHOLE ROW"

Patterns

Students may notice that once they have discovered a solution, another solution can be found by simply shifting each column over one or each row down one.

Variation

Have students compare their solutions and answer the following: Are the solutions different? Are some rotations of other solutions? Are some reflections (mirror images) of other solutions?

SQUARE COUNT

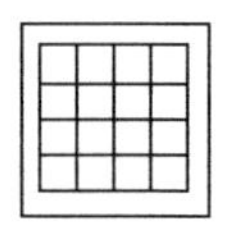

Page 18

Students explore the different ways squares can be made on two different-sized square grids. Students will need to determine a way to record their findings and may discover many numerical and visual patterns along the way.

Solutions

Puzzle 1

There are 20 possible squares that can be made on the 4 x 4 grid, as shown below. The number below each grid indicates the number of squares of that size that can be made in various positions on the board.

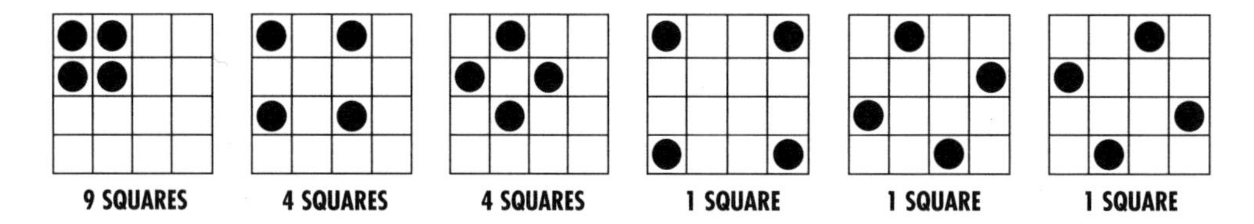

Puzzle 2

There are 50 possible squares that can be made on the 5 x 5 grid, as shown below. The number below each grid indicates the number of squares of that size that can be made in various positions on the board.

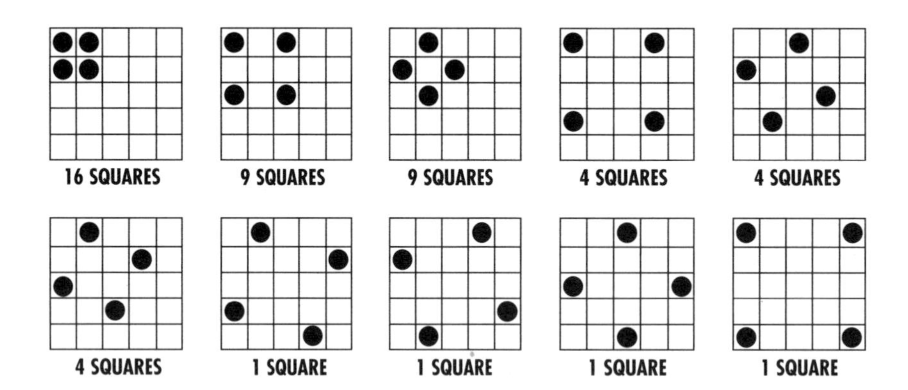

Strategies

When students find a new square, they can slide the counter over or down, one square at a time, to find all the other squares of that size that appear in the grid.

Students may note that some of the squares, such as the ones shown here, are the same size but are mirror images of each other. They cannot be duplicated by sliding the counters. Mirror images are often difficult for students to visualize. However, using an actual mirror makes it much simpler. Place the mirror right next to the grid and look into the mirror to see the reflected (or mirror) image.

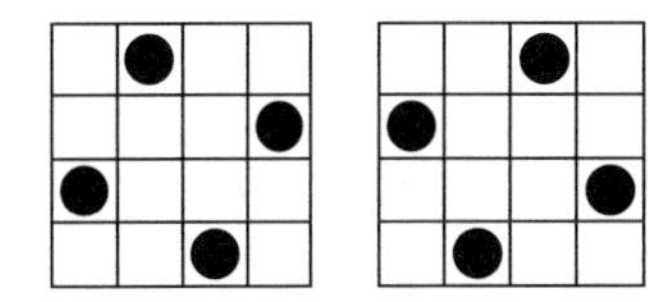

MIRROR IMAGES

Patterns

For all squares that can be found, each will appear the number of times that is equal to a *square number.* A square number is one that results when one number is multiplied by itself. Nine is a square number because 3 x 3 = 9; 16 is a square number because 4 x 4 = 16. Here are the first five square numbers in our number system: 1, 4, 9, 16, and 25.

Variations

1. Challenge students to determine the maximum number of counters that can be placed on the 4 x 4 or 5 x 5 grid without forming a square. (For the 4 x 4 square, 10 counters can be placed on the board before a square is formed. For the 5 x 5, 15 counters are possible before a square is formed. Possible solutions are shown here.)

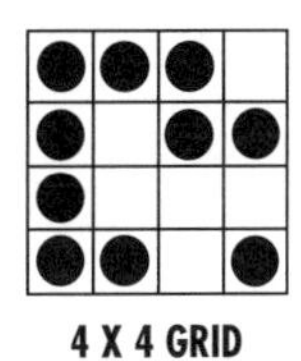

4 X 4 GRID

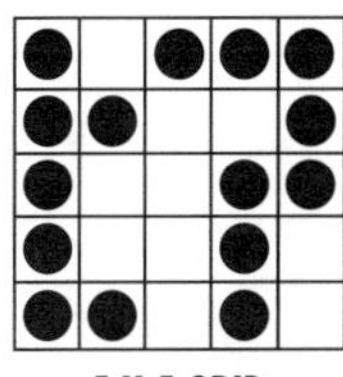

5 X 5 GRID

2. Have students create a chart picturing each different-sized square and the number of times it occurs on the 4 x 4 and 5 x 5 grids, then look for patterns. Challenge students to predict the number of times each square would occur on a 6 x 6 or 7 x 7 square grid. As mentioned earlier, each different-sized square appears the number of times equal to a square number. Knowing the square numbers, you can then predict how many times that size square will appear on the next larger grid. For example, the square shown here appears 9 times in the 4 x 4 grid and 16 times in the 5 x 5 grid. It will appear 25 times in the 6 x 6 grid and 36 times in the 7 x 7 grid. Students can test this out for themselves on the grids.

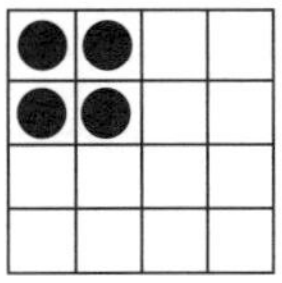

3. Challenge students to see how many triangles they can make on any of the grids using 3 counters. This will produce a wide range of new patterns.

RING TEASERS

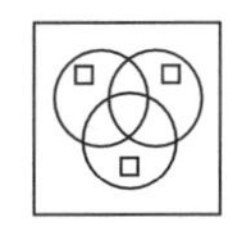

Page 20

Ring Teasers are logic problems based on Venn diagrams. For each puzzle, the total number of counters and the number of counters in each circle is given. Students must determine where to place the counters so that these numbers are true.

Do a few samples with students to be sure they understand how a counter can be placed in one, two, or three circles at the same time. Recording sheets are found on page 70.

Solutions

The two solutions shown for each puzzle are for the given puzzle and the puzzle asking for the minimum number of counters. For many, there is more than one solution.

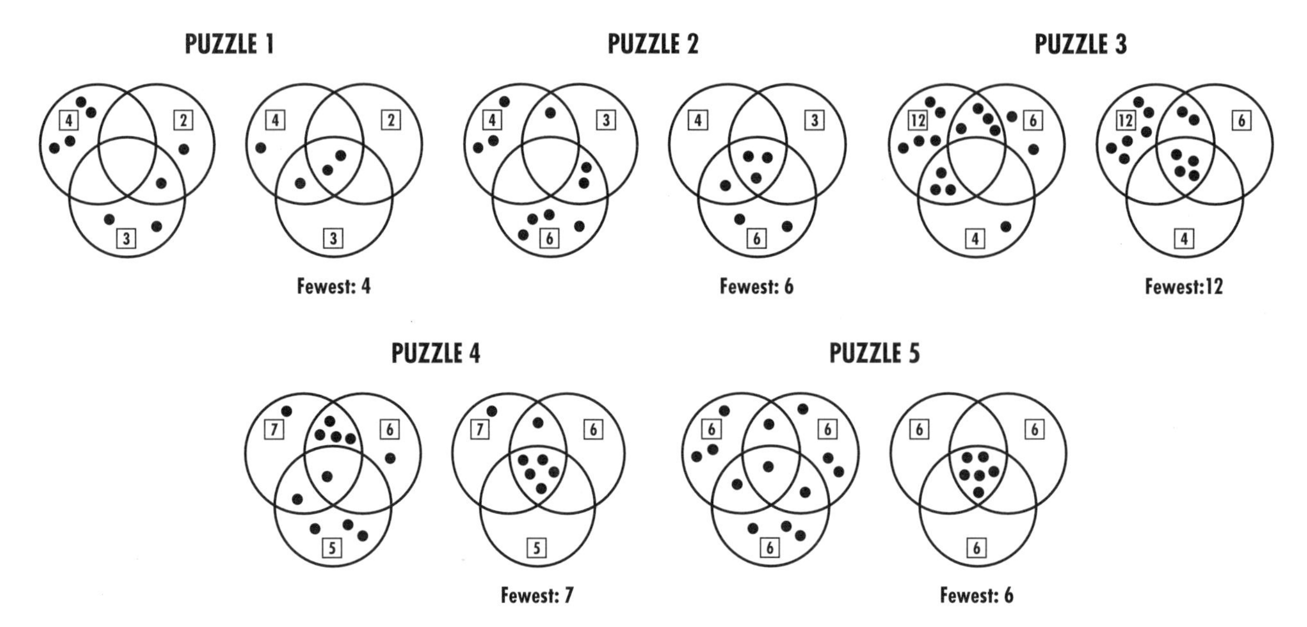

Strategy

To find the minimum number of counters for a three-ring puzzle, students discover that the minimum number of counters will be the highest number of the numbers in the squares. For example, in Puzzle 1, the highest number in a square is 4 which is the fewest number of counters needed to solve the puzzle. This is true because counters can be in two or three rings at the same time.

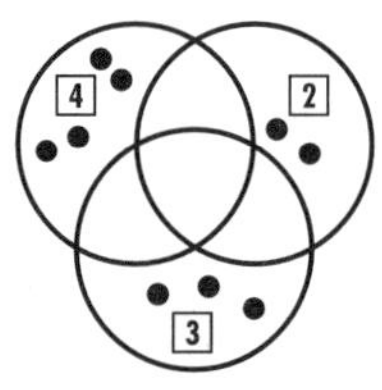

MAXIMUM NUMBER OF COUNTERS

To determine the maximum number of counters for each puzzle, students discover that they can simply place the number of counters equal to the number in the square in the outermost part of the circle. Each group of counters is in only one circle, as shown above.

For each puzzle, many solutions are possible. The varying number of counters that can be used to solve the puzzle will range from the minimum number to the maximum number. For example, for Puzzle 1, the minimum number of counters is 4 and the maximum is 9. Therefore, six different solutions are possible, using either 4, 5, 6, 7, 8, or 9 counters.

Variations

Use five rings to create Ring Teaser problems for students to solve.

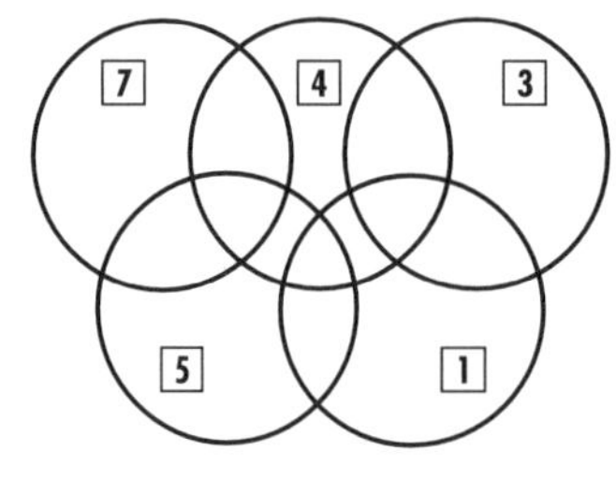

FIVE-RING TEASER

ODD EVEN ACTIVITIES

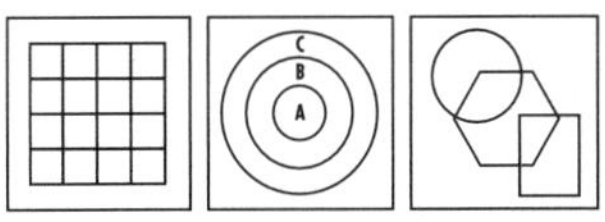

Pages 22, 24, 26

Students are challenged to arrange a number of counters on differing arrays of shapes so that a specific number of counters is placed in each array. For the circles and shapes activities, discuss beforehand how 1 counter can be placed in more than one shape at once. Recording sheets are found on pages 71–73.

ODD EVEN SQUARES

Solutions

Two solutions are shown for each puzzle, although more solutions are possible.

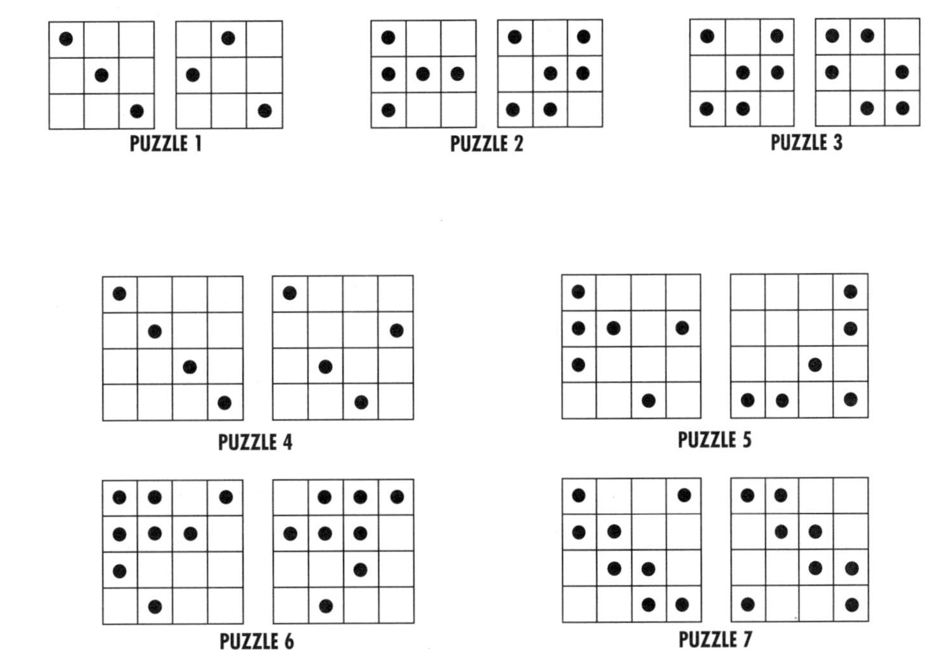

Four counters can be placed on a 3 x 3 grid so that each column and row contains an even number of counters. Students often forget that zero is considered an even number. One solution is shown here:

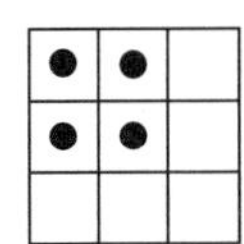

ODD EVEN CIRCLES

Solutions

Two solutions are shown for each puzzle, although more solutions are possible.

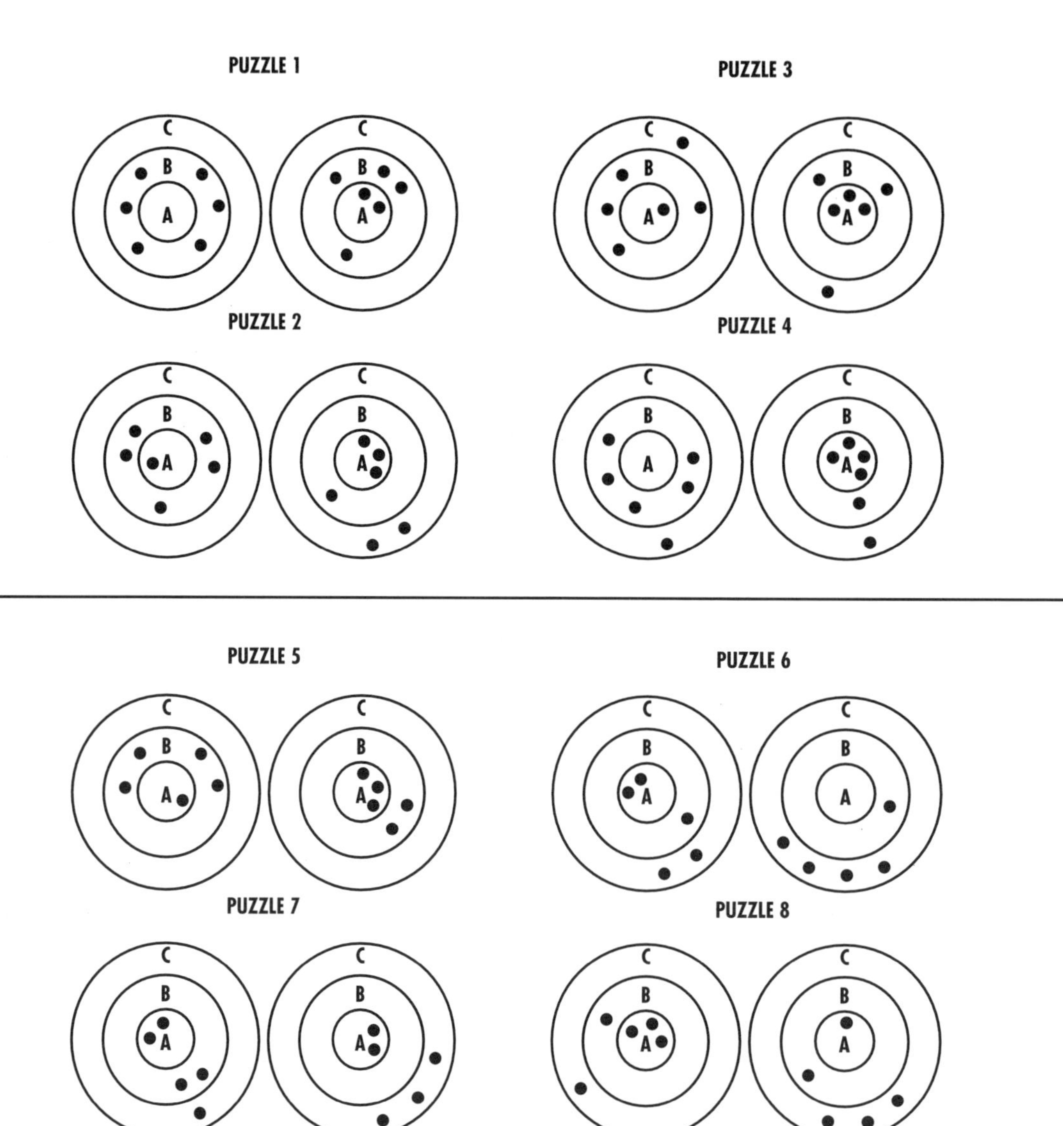

The challenge to place 7 counters in the circles so that there is an odd number in A, an odd number in B, and an even number in C is impossible to solve. The number of counters in Circle C is always the total number of counters, which in this case is 7.

ODD EVEN SHAPES

Solutions

Two solutions, using 3 counters, are shown for each puzzle, although more solutions are possible.

Two solutions, using 7 counters, are shown for each puzzle, although more solutions are possible.

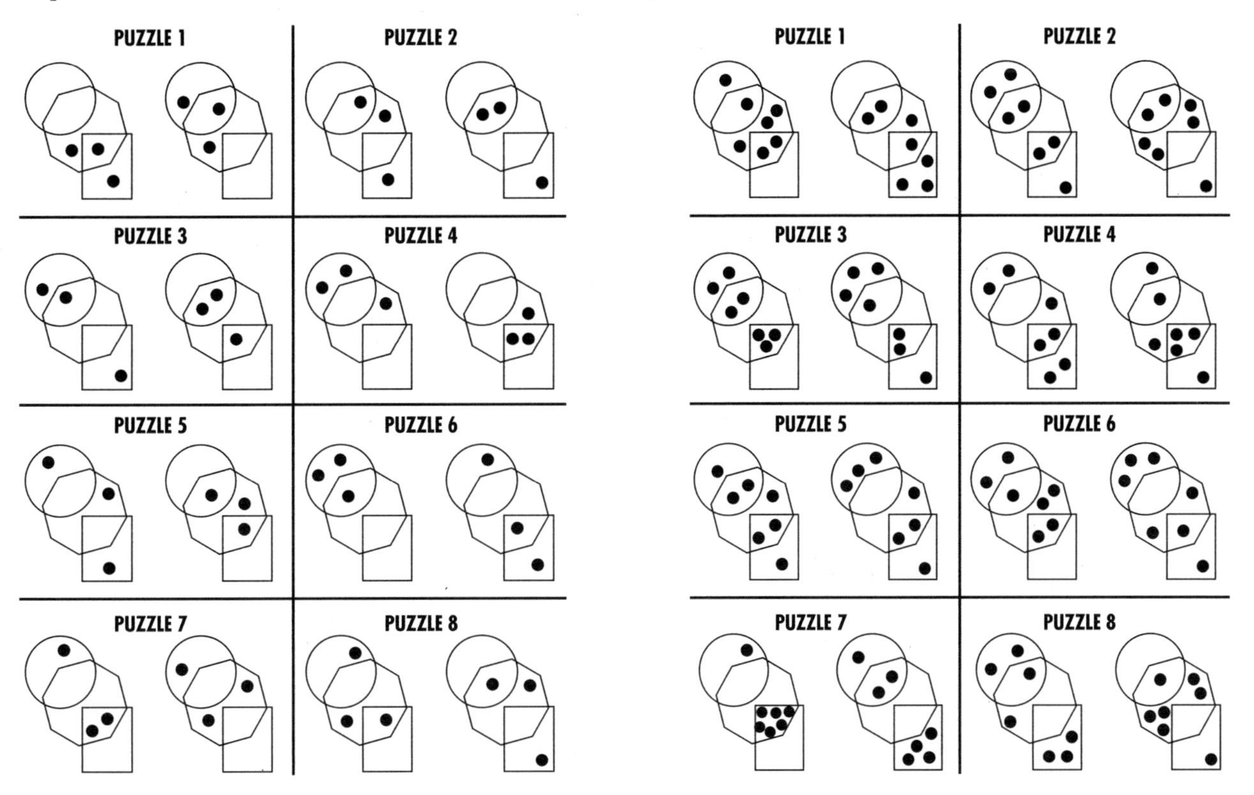

Solutions can be found for most numbers of counters for all eight puzzles. However, if using one counter, Puzzles 1, 5, and 8 are impossible to solve. A key to solving these puzzles with small numbers of counters is to remember that zero is an even number, so leaving a shape empty keeps it even.

Variations and Patterns

Challenge students to work in small groups to uncover all the possible solutions to a single puzzle in any of the games. Ask students how they organized their work and how they decided that they had found all the solutions. Some of the Odd Even Squares and Odd Even Shapes puzzles have more than 20 possible solutions, although the number of solutions for Odd Even Circles and for Odd Even Shapes (Puzzles 1–4) are generally under 10.

Finding and recording all the solutions can be very challenging for some students. If students are finding all the solutions, encourage them to work in small groups and look for patterns that will help them decide if they have uncovered all the solutions. In Odd Even Squares, for example, there are 24 possible solutions for Puzzle 4. Some students may note that some solutions are *rotations* or *reflections* of other answers. A rotation is a solution that has been turned, as shown below. A reflection is one that has been flipped or is the mirror image of the solution, as shown here.

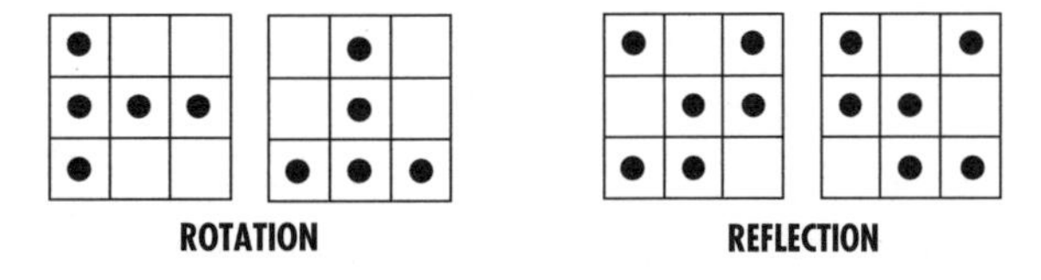

Students can decide among themselves if these 24 solutions count as unique and different solutions. Some students may decide that rotated solutions do not count as unique because you can simply turn the board and you will see the same solution. However, to see a reflected solution, you would need a mirror, so students may decide that each reflected solution counts as a new and unique one.

Some students will draw out all 24 solutions while others may see patterns along the way that enable them to predict the number of solutions.

Another pattern students may observe while solving this problem is as follows: There are four possible squares in the first column where 1 counter can be placed. When a counter is placed in one of these four places, there are then only three available squares for a counter in the second column. If a counter is placed in one of these three squares, there are then only two available squares in the third column to place a counter. If a counter is placed in one of these two squares, there is then only one available square in the fourth column to place a counter. The number of options can be multiplied to find the total number of solutions: 4 x 3 x 2 x 1 = 24.

In Odd Even Circles, students often organize their answers to help determine if they have found all the solutions. For example, if students are finding all the solutions for Puzzle 5, they may draw their solutions in order of the number of counters placed in circle A, beginning with 1 counter. This is shown below. This same information can be displayed in a chart, as shown.

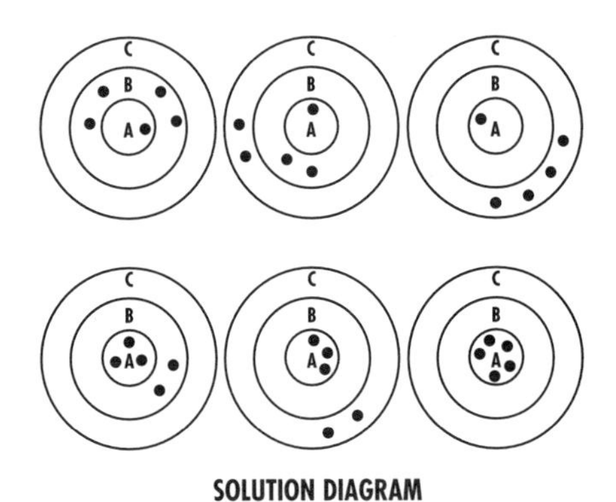

SOLUTION DIAGRAM

A	1	1	1	3	3	5
B	4	2	0	2	0	0
C	0	2	4	0	2	0

SOLUTION CHART

When using 7 counters to solve Odd Even Shapes Puzzle 1, students can find more than 20 solutions. If students are interested in finding all the solutions for a particular number of counters, using a smaller number of counters leads to a more manageable investigation.

SWITCHING PLACES

Page 28

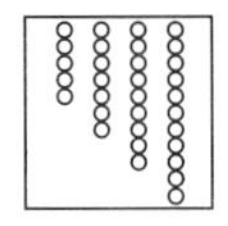

Switching Places is a classic logic puzzle requiring students to determine the fewest number of moves needed for two sets of counters to switch places. This activity provides a nice format for exploring or introducing formulas because there is a formula to determine the fewest number of moves.

Solutions

To accurately find a formula, students will need to have the correct answers for the fewest number of moves. Discuss these solutions before students investigate the last part of the Working Together section.

This chart shows the minimum number of moves for each of the game boards.

# OF RED COUNTERS	# OF BLUE COUNTERS	FEWEST MOVES
1	1	3
2	2	8
3	3	15
4	4	24
5	5	35

SOLUTION CHART

Patterns

Students may have different ways to express the pattern or rule they discovered in the data. Here is the formula for determining the fewest number of moves:

Let x = half of the total number of counters.

Then $x \cdot (x + 2)$ equals the minimum number of moves.

Students will think of many different ways to record their moves. Two ways are described here. In the first way, a picture shows the location of counters on subsequent moves. In the second, symbols represent each move.

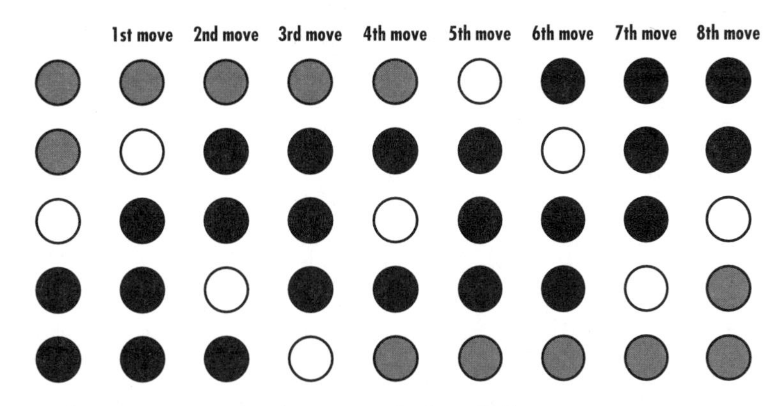

The sequence of moves can also be recorded as follows:

1R 2B 2R 2B 1R

This means that first 1 red was moved, then 2 blues were moved, then 2 reds, and so on. A similar pattern occurs for the 6-counter problem. Here is the sequence of moves:

1R 2B 3R 3B 3R 2B 1R

This is the sequence for the 8-counter problem:

1R 2B 3R 4B 4R 4B 3R 2B 1R

Ask students to predict the sequences for other Switching Places puzzles.

Variation

Have students try a variation called the Tower of Hanoi.

There are three non-intersecting circles. On one circle, 3 counters are stacked, red on the bottom, blue in the middle, and green on top. Students must move them, one at a time, so that they are stacked in the same order on one of the other circles. Challenge students to determine the fewest number of moves needed to move the counters to a different circle. The counters can be moved according to the following rules:

- Only 1 counter can be moved at a time.
- A counter covered by another counter cannot be moved.
- The blue counter can never be placed on top of the green counter, and the red counter can never be placed on top of the blue or green counter.

The Tower of Hanoi problem can be solved in seven moves. Call the circles 1, 2, and 3. Each counter will be identified by its color.

To start: R1, B1, G1 (This means that all the disks are stacked on Circle 1.)
Move 1: R1, B1, G2 (The green counter is moved to Circle 2.)
Move 2: R1, B3, G2
Move 3: R1, B3, G3
Move 4: R2, B3, G3
Move 5: R2, B3, G1
Move 6: R2, B2, G1
Move 7: R2, B2, G2

CRISS-CROSS

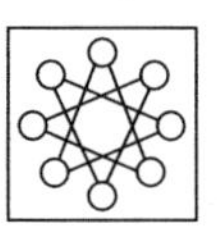

Page 30

Criss-Cross is a seemingly simple puzzle that will keep students thinking for a while. It can be played with a partner or as a solitaire game.

Strategy

To win Criss-Cross, follow this strategy: Whenever you place a counter on a circle to be moved to its final destination, that circle should also be the final destination of the next counter to be placed. For example, place a counter on Circle A and slide it to its final destination. Place the next counter on Circle B so that its final destination is Circle A. The next counter you place should have Circle B as its final destination.

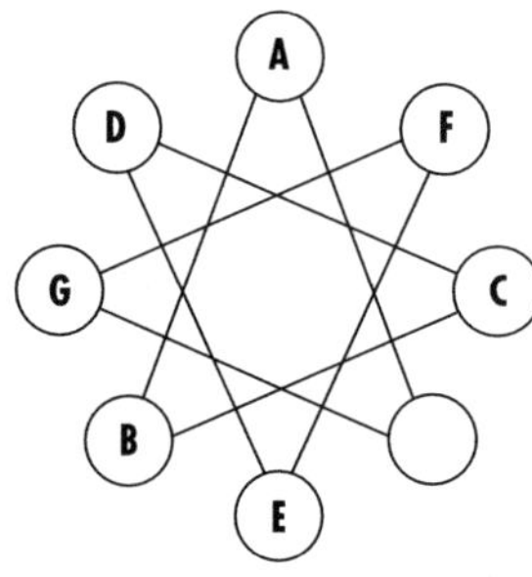

CRISS-CROSS SOLUTION

Variations

For the variations listed below, note that if students have discovered the strategy, these variations are often simple to solve. If students have not discovered the strategy, then these variations can often help students work toward solving the puzzle.

1. Use 3 red counters and 4 blue counters. Following the same rules, challenge students to place the 7 counters on the board so that they follow a red, blue, red, blue, and so forth pattern.
2. Use 3 red counters and 3 blue counters. Following the same rules, challenge students to place the 6 counters on the board so that the 3 counters of each color are together.
3. Use 7 different-colored counters. Following the same rules, challenge students to place the 7 counters on the board so that a specific order results, such as a chain that goes red, blue, yellow, green, black, white, orange.

CRAZY CUPS

Page 32

This is a classic logic problem that examines the idea of sets and how one set can be a subset of another group. Students may not realize that the cups can be placed inside of one another to find solutions. If, after given time to generate solutions, students do not discover an answer involving stacking the cups, show students how one cup can be placed inside another and how the counters in the first cup are then included in the second cup.

Stacking, see-through plastic cups work best for this activity as students will have to place one cup inside another.

Solutions

Puzzle 1 There are 25 solutions. In four of the solutions the cups are separate, as shown.

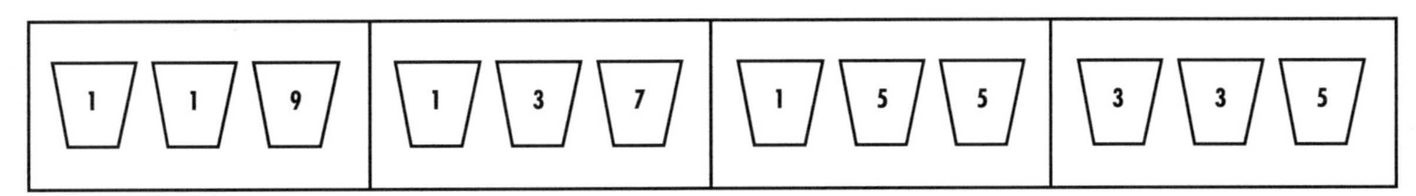

SOLUTIONS USING SEPARATE CUPS

For the remaining 21 of the solutions, all three cups are stacked inside one another. Cup A is inside Cup B, which is inside Cup C. These 21 solutions are shown in the chart below.

CUPS	NUMBER OF COUNTERS IN EACH CUP																				
A	1	1	1	1	1	1	3	3	3	3	3	5	5	5	5	7	7	7	9	9	11
B	0	2	4	6	8	10	0	2	4	6	8	0	2	4	6	0	2	4	0	2	0
C	10	8	6	4	2	0	8	6	4	2	0	6	4	2	0	4	2	0	2	0	0

SOLUTIONS USING STACKED CUPS

Puzzle 2 There are 15 solutions to this problem, as illustrated in the chart.

1 / 0, 9	1 / 2, 7	1 / 4, 5	1 / 6, 3	1 / 8, 1
3 / 0, 7	3 / 2, 5	3 / 4, 3	3 / 6, 1	5 / 0, 5
5 / 2, 3	5 / 4, 1	7 / 0, 3	7 / 2, 1	9 / 0, 1

The solutions can also be presented in Venn diagram form.

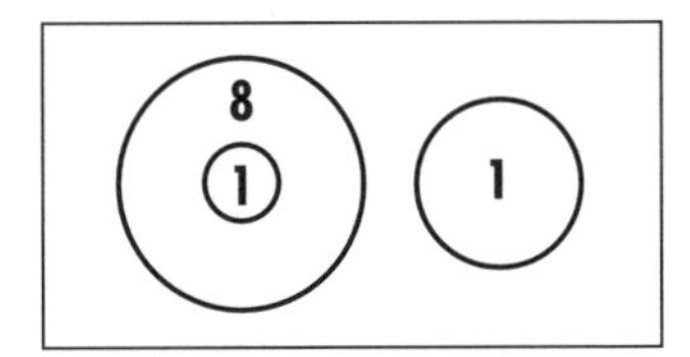

SOLUTION IN VENN DIAGRAM

Patterns

It is often challenging for students to determine if they have found all the solutions. One method that helps is to organize the answers into a chart like the one in Puzzle 1 which shows the solutions using stacked cups. If answers are placed in an organized fashion, number patterns become apparent. For example, if you look across the chart for Cup A, you see that the odd numbers 1, 3, 5, 7, 9, and 11 appear in order. You can also see that there are six solutions with a 1, five solutions with a 3, and so on. Each time there is one less solution. Noticing these patterns can help students determine if they have found all the solutions. For example, a student who has found only three solutions with 5 counters in Cup A might wonder if another exists.

BAD APPLE

Page 34

Bad Apple is a two-person strategy game, like Nim (page 38) and Circle Game (page 36), that involves removing counters to win. While playing, many students will realize that if they leave their opponent with 5 counters, they can win the game. It is harder to have students take another step back to determine the moves that will always leave their opponent with 5 counters.

Strategy

To win the game with 13 counters, be the second player and follow this strategy: Play in groups of 4—meaning that whatever your opponent takes, you take the number of counters that makes the total taken for that round equal 4. For example, if your opponent goes first and takes 3 counters, you take 1 counter. This works because there are 12 counters that you want to take; it's the thirteenth that you don't want.

If the game is changed to 15 counters, the strategy changes as well. In this version, if you go first, you can always win by taking 2 counters on your first move, which leaves your opponent with 13 counters. Now you can follow the strategy described above for 13 counters.

Variations

1. Vary the number of counters used or the number of counters that can be taken off. For example, use 11 counters and allow players to remove 1 or 2 counters at their turn. Discuss how this changes the strategy.
2. Change the goal of the game by making the player who takes the last counter the winner. Challenge students to discover a winning strategy for this version. (To guarantee that you take the last counter in the 13-counter game, be the first player and take 1 counter. Then play in groups of four. If your opponent takes 3 counters, you take 1. If your opponent takes 2 counters, you take 2. This will assure you of taking the last counter.)

CIRCLE GAME

Page 36

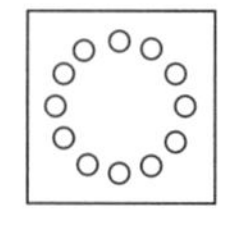

The Circle Game is a two-person strategy game, like Nim (page 38) and Bad Apple (page 34), that involves removing counters to win. The game is easy for students to learn, but discovering a winning strategy is challenging.

Strategy

To guarantee a win each time, you must be the second player and follow this strategy: On the first turn, Player 1 will take 1 or 2 counters from anywhere in the circle. This will leave a gap in the circle. Player 2 must remove 1 or 2 counters opposite the gap so that Player 1 is left with two equal chains of counters. For example, if Player 1 takes 1 counter as shown, Player 2 takes 1 counter so that there are two chains of 4 counters each.

From then on, Player 2 must always remove counters so that regardless of what Player 1 takes, there are always equal chains after Player 2's move. In this fashion, Player 2 will always win. This strategy works for the 10-counter and 13-counter versions.

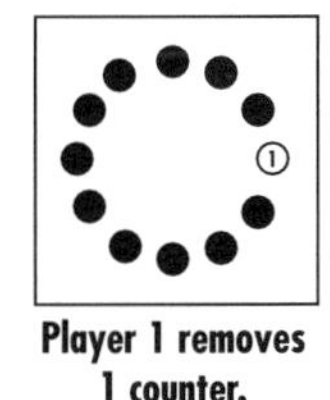

Player 1 removes 1 counter.

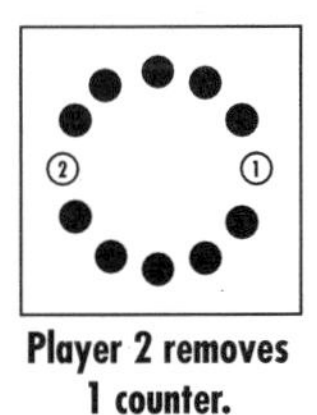

Player 2 removes 1 counter.

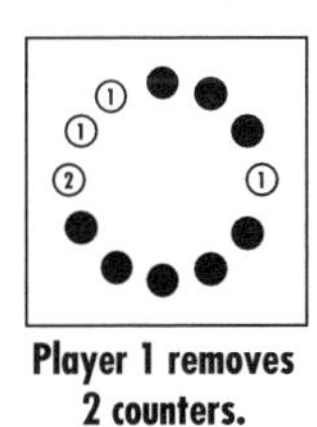

Player 1 removes 2 counters.

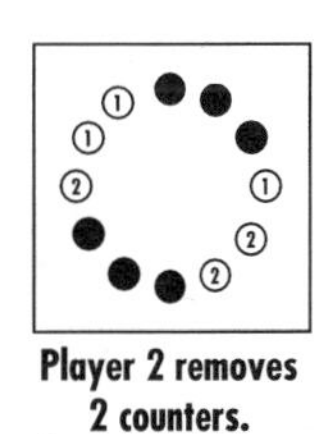

Player 2 removes 2 counters.

Variation

Play the game using any number of counters. The same strategy works for all games.

NIM

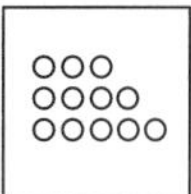

Page 34

Nim is an old and classic two-person game that is easy to learn and is jam-packed with mathematics.

Strategy

A winning strategy for this version of Nim is to go first and take 2 counters from the top row. From then on, either take counters so that your move leaves 1 counter in one row, 2 counters in another row, and 3 counters in another row. Or take counters so that your move leaves exactly two rows with the same number of counters in each of the two rows.

The mathematics behind the game involves the base-two number system. Base two can be thought of as the doubling pattern: 1, 2, 4, 8, 16, and so forth. To write a number in base-two notation, use the numbers in the doubling pattern to add up to your number—with the restriction that a number can only be used once. For example, 7 equals 4 + 2 + 1, and is written in base two as 111_2. Ten equals 8 + 2, and is written as 1010_2. The zeros indicate that 4 and 1 were not used, and the small 2 at the end of the number indicates the base of the number system.

16	8	4	2	1	
		1	1	1	= 7
	1	0	1	0	= 10

To examine the game strategy, record the counters in each row in base-two notation, or use a chart as shown below. The goal is to always keep the totals even. For example, the board at the beginning of the game can be counted as shown below in Chart 1. Note that the Totals row and the Counters column are not recorded in base two.

	Counters	4s	2s	1s
Top Row	3		1	1
Middle Row	4	1	0	0
Bottom Row	5	1	0	1
Totals	12	2	1	2

CHART 1: NIM GAME BOARD AT THE BEGINNING OF THE GAME

By looking at the Totals row, you can see that there is an odd number of twos. To even off the number, remove 2 counters from the top row. This leaves an even number of counters—look in the Totals row for Chart 2 in which all the numbers are even. Each move can then be analyzed in this fashion.

	Counters	4s	2s	1s
Top Row	1			1
Middle Row	4	1	0	0
Bottom Row	5	1	0	1
Totals	10	2	0	2

CHART 2: NIM GAME BOARD AFTER REMOVING 2 COUNTERS FROM THE FIRST ROW

The strategy, therefore, is to play first and keep the board even, as counted in base two. If you do this, you will win every time. This base-two counting method is helpful for determining strategies when the number of counters and rows is altered.

Variations

1. Alter the game by including more rows in the game or more counters in the rows, such as a game with rows of 6, 7, and 8 counters. When the number of counters and rows is altered, the winning strategy depends on the set up of the counters. To determine the winning strategy you must determine if the game board is an even or odd board as counted in base two (described above in the Strategy section). If the board at the start of the game is even, as counted in base two, then be the second player and always make the board even after your turn. If the board at the start of the game is odd, as counted in base two, then be the first player and always make the board even after your turn.
2. Change the rules so that the player who takes the last counter loses. To win this version when playing with the original 12-counter, three-row format, play by following the original strategy. However, at some point in the game, you must either leave your opponent with an odd number of single-counter rows or leave your opponent with two rows, each containing the same number of counters. This will guarantee that your opponent will take the last counter.

LINE UP

Page 40

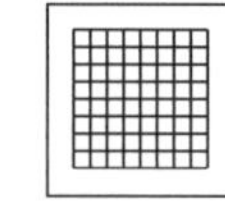

Line Up is a game in which players must carefully examine the board as they place their counters and try to make a row, column, or diagonal of 3 or 4 counters. Some students may find a straight-edge helpful for seeing the diagonal lines. This game provides a nice follow-up to the ideas explored in Tight Fit (page 8).

Solution and strategy

This game is like a larger version of Tic-Tac-Toe. A winning strategy is to be the first player and make your first move anywhere except in the outer row of squares. For your second move, place the counter adjacent to the first counter in such a position so that you have two ways of winning.

In the example shown here, Black is Player 1. White cannot block Black because Black has two ways of making a row of 3: placing a counter either above or below the row of 2 counters already on the board. Following this strategy, the first player should always be able to win on the third move.

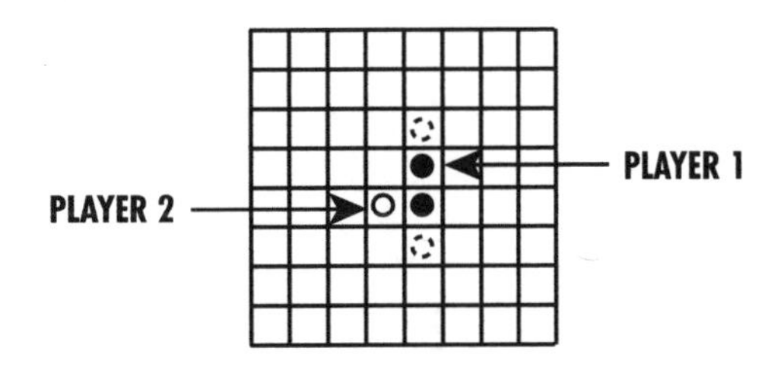

Using 4 counters is more challenging. Finding a winning strategy is difficult, but if you can get 3 counters in a row with open spaces at both ends before your opponent does, you can always win.

Variations

Have the students try to force their opponents to make the first row, column, or diagonal of 3 counters. Play this game on a 4 x 4 and then a 5 x 5 game board. (Creating a foolproof winning strategy for these versions is difficult. The games often come down to the last few squares where any move a player makes results in a row of 3.)

HEX

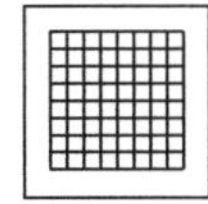

Page 42

Hex is a strategy game for two players. The game is simple to learn but difficult to master. The original game, created by a Danish mathematician named Piet Hein, was designed for a board with 11 hexagons on a side. But it is very difficult to develop a winning strategy on a large board. Smaller game boards make the game and its strategy accessible to students.

Strategy

There is a winning strategy for each of the four game boards.

For the 2 x 2 game board: Go first, as the first player always wins on the second move.

For the 3 x 3 game board: Go first and put a counter in the center hexagon. Because the center hexagon is connected to two other hexagons on each side, the second player cannot block the first player. The first player can then win on the third move.

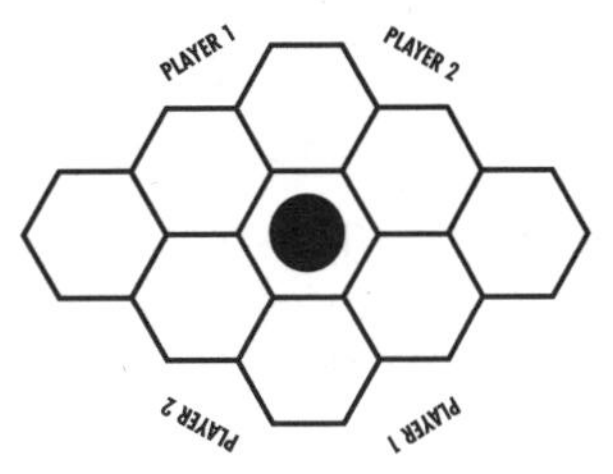

FIRST MOVE ON 3 X 3 GAME BOARD

For the 4 x 4 game board: Go first and put a counter in any of the four center hexagons, as shown here. The first player can then win on the fourth round of moves.

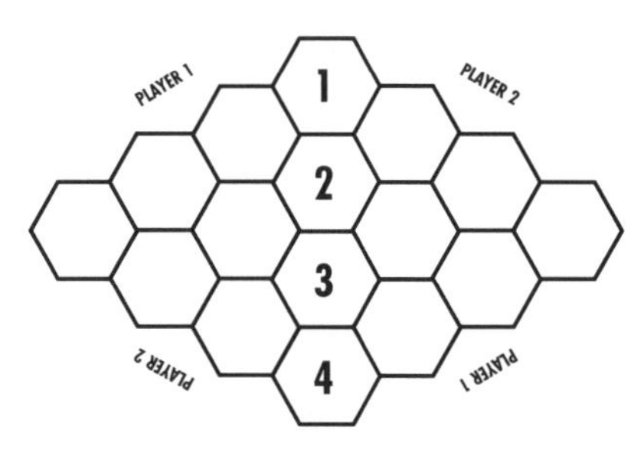

FIRST MOVE ON 4 X 4 GAME BOARD

For the 5 x 5 game board: Go first and put a counter in any of the five center hexagons, as shown here. The first player can then win on the fifth round of moves.

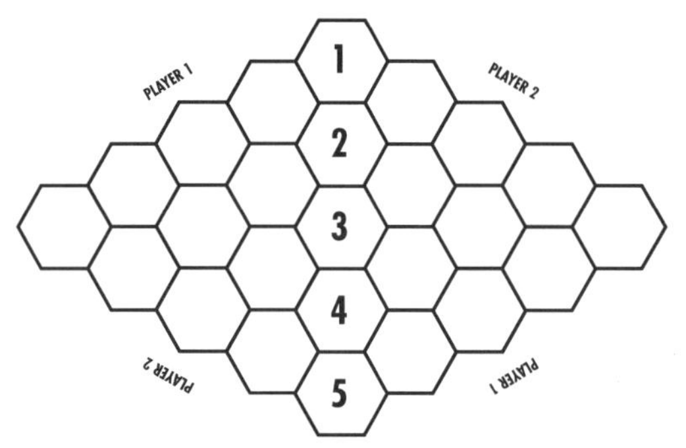

FIRST MOVE ON 5 X 5 GAME BOARD

Variation

Play Hex on larger game boards. One way to create a larger board is to photocopy one of the game boards onto a heavier stock of paper. Cut out one hexagon and use this for a tracing template.

RECORDING SHEETS

CIRCLE MYSTERIES PART I RECORDING SHEET

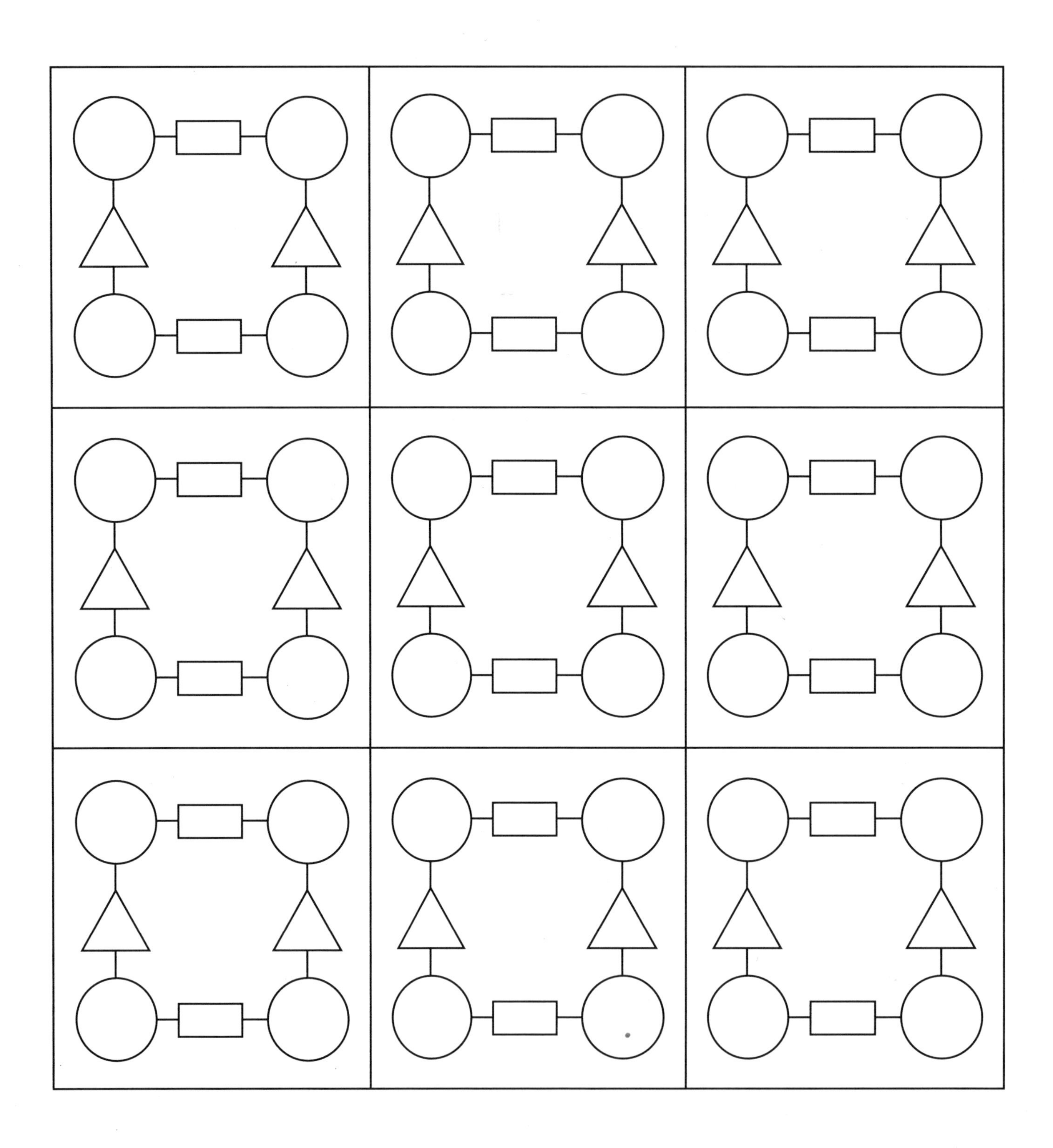

CIRCLE MYSTERIES PART II RECORDING SHEET

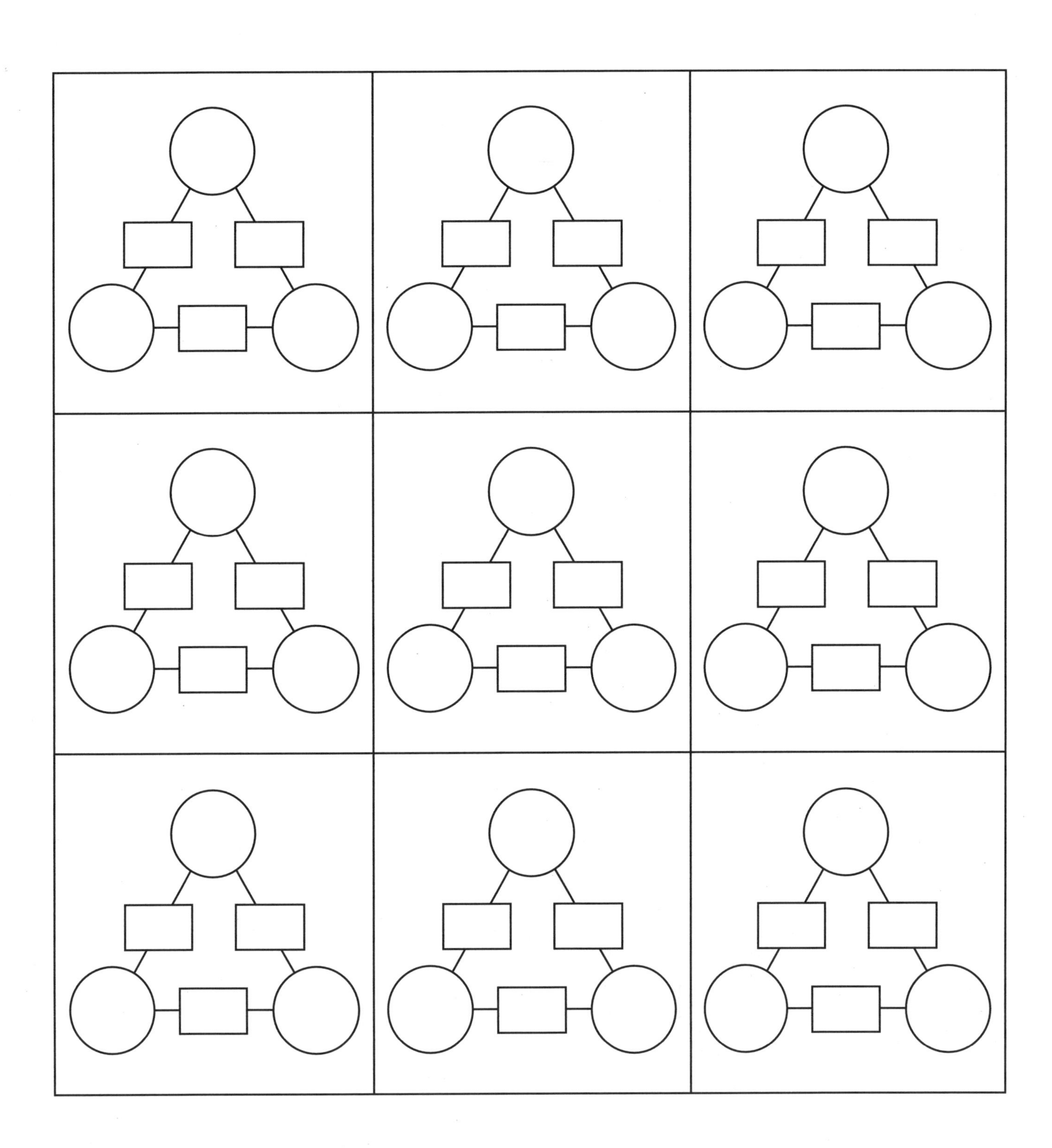

RING TEASERS RECORDING SHEET

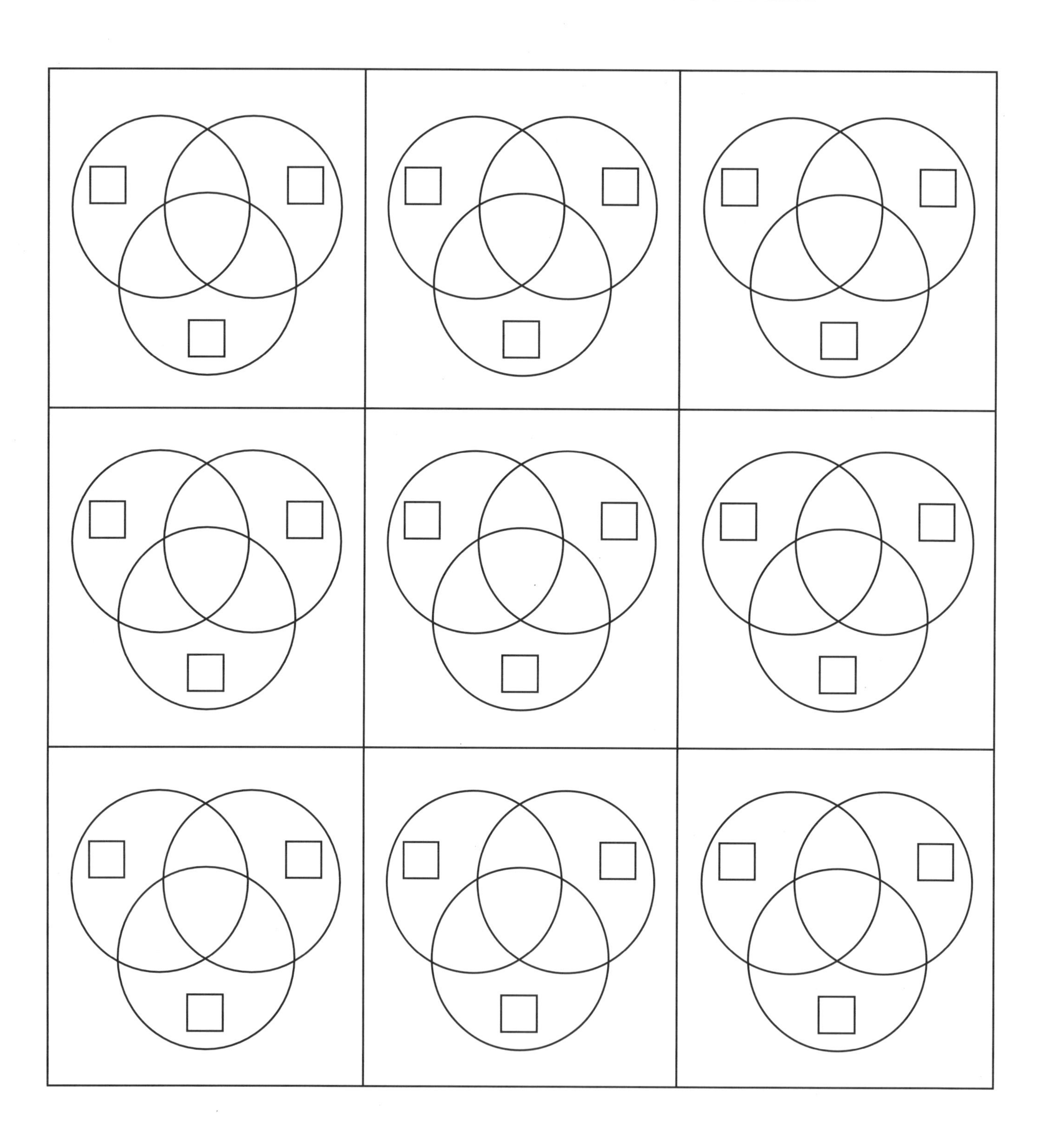

ODD EVEN SQUARES RECORDING SHEET

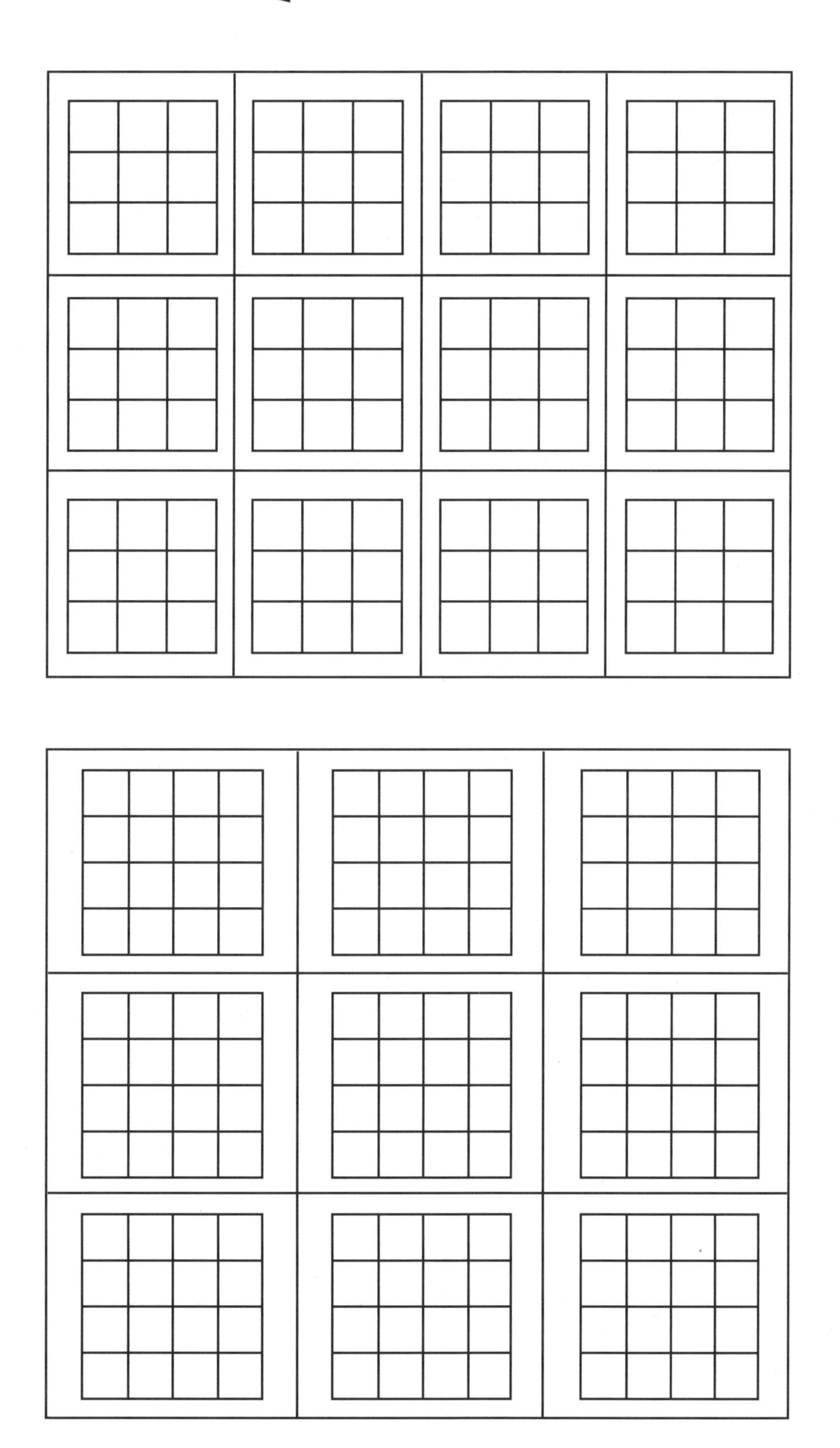

ODD EVEN CIRCLES RECORDING SHEET

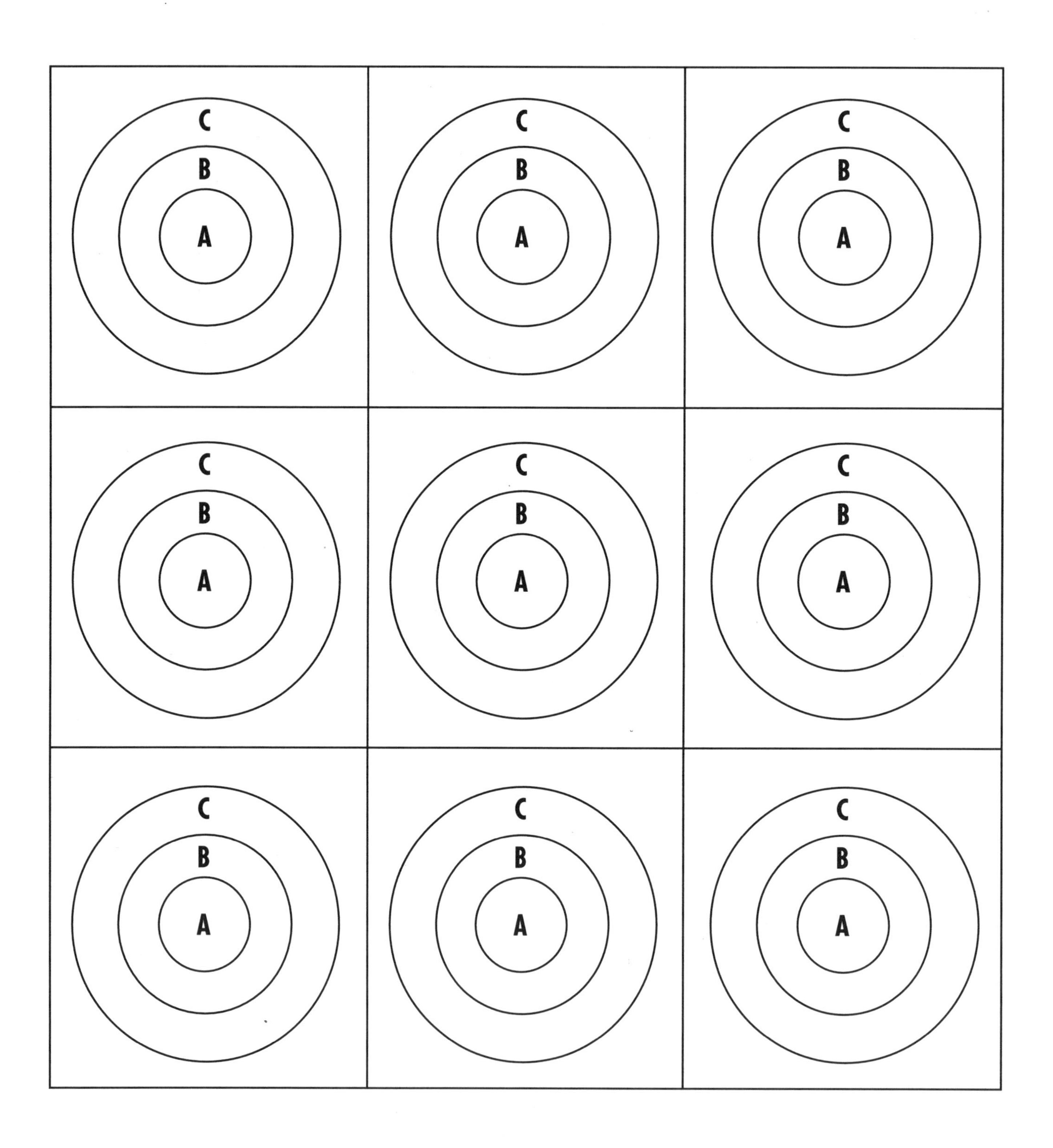

ODD EVEN SHAPES RECORDING SHEET

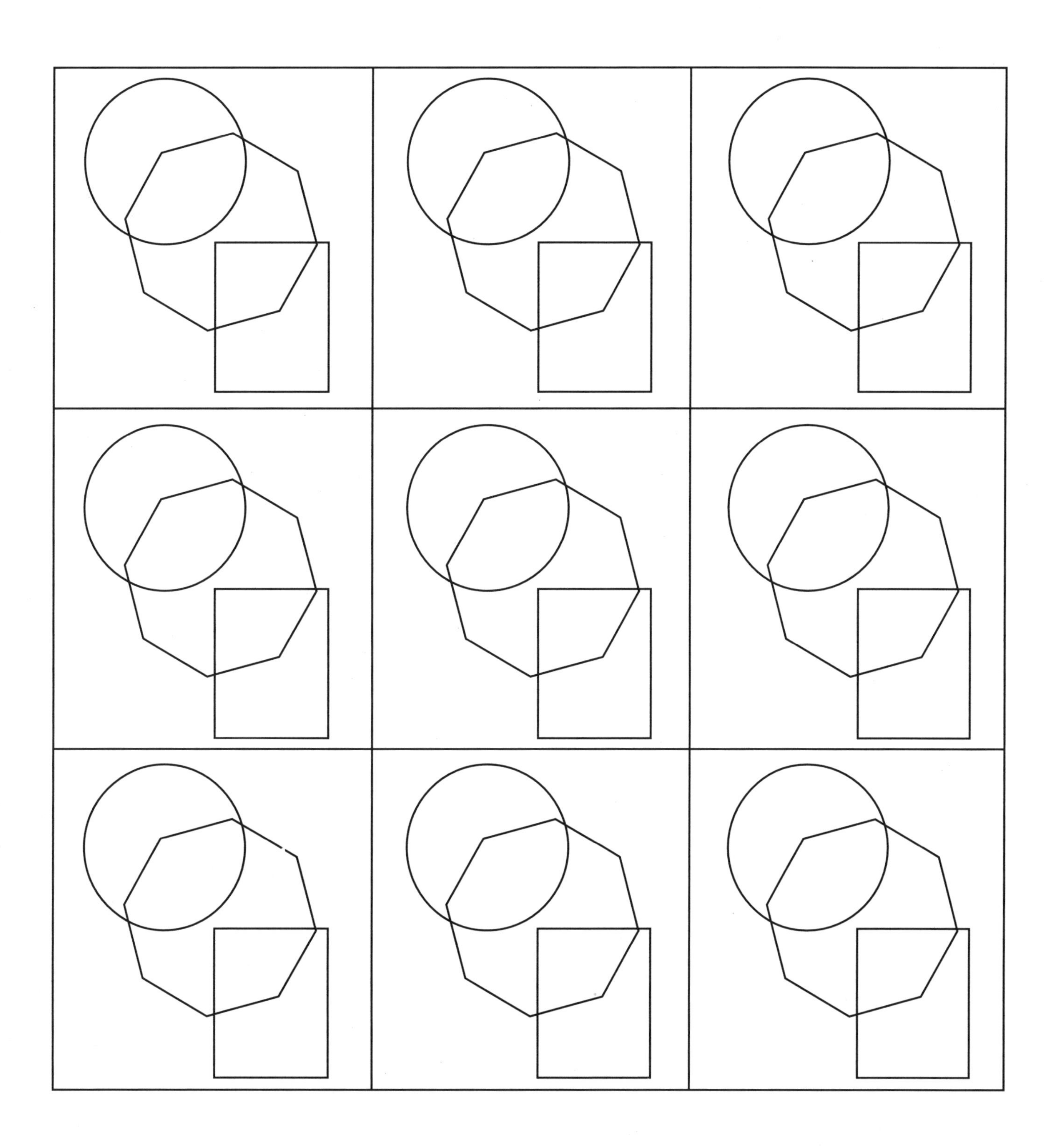